HOW TO WRITE UP YOUR GEOGRAPHY DISSERTATION

This step-by-step guide to the writing journey of a geography dissertation provides guidance and reassurance on the foundational aspects of the dissertation. The guide will also support students in focussing their time and energy on the substantive components and concerns of their individual work.

It is no surprise that the dissertation is often a cause of stress and anxiety for students. Much of this is linked to students' struggles around questions and concerns about how to structure the work, what content should go in which chapter, the purpose, structure, and content of each chapter, how to meet the word limit, and other similar considerations. This text addresses in detail these common areas of concern, covering not only guidance on how to develop a writing strategy that suits the individual but also guidance on writing style to help students convey their breadth and depth of knowledge within the word limit of a dissertation, as well as providing a chapter-by-chapter guidance on purpose and content.

Written in a clear, accessible, and engaging style, this detailed guide will empower students to understand and deliver a successful dissertation. An invaluable resource for undergraduates, it will also reduce the burden on supervisors from commonly asked questions. In addition, the author presents advice on meeting the expectations of markers and how to use the dissertation in the job market.

Daniel Hammett is a senior lecturer in the School of Geography and Planning, University of Sheffield, UK, and a senior research associate in the Department of Geography, Environmental Management and Energy Studies, University of Johannesburg, South Africa. He has over 20 years of national and international field research and field class teaching experience across numerous national contexts.

His research focusses on questions of citizenship, nationalism, and geopolitics, and his research has been published in journals, including *Progress in Human Geography*, *Political Geography*, and *Transactions of the Institute of British Geography*, as well as recent books with Routledge (*Global Development: The Basics*) and Bristol University Press (*Humour and Politics in Africa: Beyond Resistance* – with Laura S. Martin and Izuu Nwankwọ).

HOW TO WRITE UP YOUR GEOGRAPHY DISSERTATION

A STEP-BY-STEP GUIDE

Daniel Hammett

Routledge
Taylor & Francis Group
LONDON AND NEW YORK

Designed cover image: Getty Images

First published 2026
by Routledge
4 Park Square, Milton Park, Abingdon, Oxon OX14 4RN

and by Routledge
605 Third Avenue, New York, NY 10158

Routledge is an imprint of the Taylor & Francis Group, an informa business

British Library Cataloguing-in-Publication Data
A catalogue record for this book is available from the British Library

ISBN: 978-1-032-95326-7 (hbk)
ISBN: 978-1-032-95324-3 (pbk)
ISBN: 978-1-003-58431-5 (ebk)

DOI: 10.4324/9781003584315

Typeset in Bembo
by SPi Technologies India Pvt Ltd (Straive)

CONTENTS

FIGURES

TABLES

ACKNOWLEDGEMENTS

This book has only been possible because of the guidance received from colleagues and the experience gained through supervising countless undergraduate and postgraduate dissertations over many years. While it is impossible to mention everyone individually, the mentoring and support from colleagues Chasca Twyman and Charles Pattie were particularly formative during my early career as a lecturer (and the late Lynn Staeheli prior to this). This book has also benefitted from specific conversations and reflections on dissertations and supervision approaches across different sub-disciplinary areas and methodological approaches with Luke Temple, Julie Jones, Sammia Poveda, Lucy Jackson, Richard Phillips, and Peter Jackson among others. Beyond academic support, this book has benefitted immeasurably from the support of the team at Routledge – Niamh Hitchmough and Claire Maloney in particular – as well as the input of anonymous reviewers. Finally, the constant and consistent support, love, and forbearance of my family have underpinned not only the writing of this book but also my academic career – for this I am eternally grateful.

1

THINKING ABOUT YOUR DISSERTATION

Your dissertation is *the* crucial piece of assessment in your degree programme. It will be longer and more complex than any other assignment on your course. If you have been writing 1,500- to 3,000-word essays for the past two years or more, then a dissertation of somewhere between 8,000 and 15,000 words will likely feel like an insurmountable challenge. Initially, at least, you will be worrying about where all those words will come from. By the end of the process, you will probably be wishing you have another few thousand words to work with as you wrestle with editing an over-length draft down to your departmental limit. (Recognising institutional structures vary and you may be based in a school, a department, an institute, or equivalent, for the sake of simplicity this book will use department in a generic sense to refer to all of these).

This book is designed to help you through this process, whether your dissertation is 8,000, 10,000, or 15,000 words long (or anywhere in between). Hopefully you will be reading this as you plan and write your dissertation proposal. More likely you are later in your journey and beginning the write-up phase of your dissertation. Regardless of where you are in the process when you read this book, the focus and objective remain the same: to provide step-by-step guidance on the different components of a successful geography dissertation, including strategies and advice on how to write you dissertation, and how to write well when doing so. At the core is a focus on writing for your primary audience – your examiner – although this is expanded in the closing chapters of the book to consider how you might effectively communicate your research to

DOI: 10.4324/9781003584315-1

other audiences and mobilise this work to support your employability. Before going further, however, it is useful to pause and think in more general terms about your geography dissertation.

YOUR GEOGRAPHY DISSERTATION

What is your dissertation? This may seem a banal or overly abstract question, but understanding what your dissertation is, why it is required, and for whom you are writing it provides key insights into both how to approach the work and what you need to deliver at the end of the process.

In a broad sense, your dissertation

- is an opportunity to focus on something of interest to you and to make a contribution to knowledge about this topic;
- an independent project that requires you to design, deliver, and effectively communicate a planned programme of research;
- a piece of work in which you evidence a range of technical, academic, professional, transferable, and communication skills.

Your dissertation is thus a key opportunity to develop your knowledge and understanding in detail on a specific topic, issue, or place that you find particularly interesting. In other words, it is a research project addressing a set of original and specific research aims and/or questions that provides your reader with new insights and information into the world around them. It will involve the identification of a topic of interest, development of pertinent research questions, engagement with relevant literature, and the design and execution of a research design and plan. In turn, this means you will collect and/or collate various forms of data – whether primary or secondary, qualitative or quantitative – and subject these to appropriate, rigorous analysis.

Wherever your research is focussed, on whatever set of questions within whichever sub-disciplinary branch of geography, and using any kind(s) of data, your dissertation will be worthless if you cannot effectively communicate this work to your audience. Rather than leaving the writing-up of your dissertation to the last minute or as a relative afterthought, this stage of your dissertation journey needs to be taken seriously. You may have collected the most incredible data, developed deeply insightful analysis that provides new perspectives on a topic – but if you are unable to explain what you have done, what you have found out, and how this understanding builds on existing knowledge, this work is wasted. Writing up your dissertation, therefore, is a crucial step ensuring you effectively communicate your research. On an instrumental level, it is essential for your degree outcome. To this end, your dissertation needs to clearly communicate

1. the question(s) you have asked;
2. why the question(s) matters;
3. how you have attempted to answer the question(s);
4. what you have found out;
5. what your answer to the question(s) is.

Your dissertation is, in essence, an opportunity to 'stand on the shoulders of giants' and develop new insights into geographical debates and knowledge. Nonetheless, you need to explain and evidence that your dissertation is rooted in a critical understanding of existing debates *and* based on your own research and analysis. From this, you need to demonstrate the original findings presented and how these make a 'contribution to knowledge'.

CONTRIBUTION TO KNOWLEDGE

The idea of your dissertation making a 'contribution to knowledge' appears in various guises within dissertation marking schemes, framed in the langue of demonstrating 'originality', 'innovation', 'presenting new ideas or challenging received views', or similar terminology. This may initially sound intimidating and off-putting. But what does this actually mean for your dissertation? In reality, the expectations associated with this language are much more modest than you might fear. Aptly for geographers, the key consideration here is about scale – the scale of expectations held by your examiners, your supervisors, and you. Your dissertation is not a lifetime's work, nor is it going to be the next best-selling popular geography book.

No-one is expecting your undergraduate dissertation to fundamentally change our understanding of volcanic gas emissions nor lead to a retheorising of the meaning of public space nor provide an agenda-setting argument on encouraging biodiversity in UK salt-marshes. Rather, it means you need to do something – anything – a bit differently to previous work. Your dissertation's 'contribution to knowledge' will be a very modest addition to the gradual accretion of understanding that is the purpose of academia. For an undergraduate dissertation your examiners are looking for you to ask a specific geographical question and, through a detailed exploration, effectively answer this query to an appropriate degree. In practice this might mean engaging with a niche oversight in existing literature, or offering a new case study to critically reflect on an existing concept, or asking a new question of an existing dataset.

Your originality or innovation might therefore be in providing a comparison of media coverage of an event (or events) over time and between sources, or in measuring and analysing water flow and quality in a particular river, or by modelling changing flood risk within a specific area, or exploring the impacts of a specific infrastructural or housing development on a local community or environment. Your ability to articulate how this work presents new ideas or insights or challenges received wisdom relates to your critical engagement with and synthesis of existing literature, and your mobilisation of this to frame and inform your analysis and argument. This may be realised through a focus on a topic or issue close to home or overseas, through the collection and analysis of primary data or the analysis of existing qualitative or quantitative secondary data.

There are two key points here. The first is to ensure you can identify and articulate your contribution to knowledge, no matter how modest. The second is to be realistic about what your contribution – your key finding – is, and not to overclaim this. Any claim to providing 'agenda-setting insights' or 'paradigm-altering findings' will be met with scepticism. Your examiner will be more impressed if you recognise and reflect on the limitations to your project and effectively articulate the modest contribution to knowledge being made than try to overclaim. Not only does overclaiming indicate a degree of hubris but simultaneously calls into question your understanding of the field of research and your ability to think critically.

The rest of this book aims to help you avoid this – and other – common mistakes while providing detailed guidance on how to write your dissertation to ensure you clearly communicate your 'red thread' (for more details on this, see Chapter 4 – *Good writing*) and develop an effective structure and approach for your dissertation. To do this it is first important to understand – and be able to articulate – why your dissertation matters.

(WHY) DOES MY DISSERTATION MATTER?

Your dissertation matters on several levels, some of which need clearly articulating to your reader. Others are more instrumental drivers and motivators that do not need to be explicitly communicated in your writing.

As a geography student, we can assume you are interested in and want to know more about the world around you. As Kimberley Peters (2017: 21) outlines, being a geographer implies a concern with and interest in writing and producing knowledge about the world – that you are not only writing about the world but 'writing worlds'. Your dissertation is a moment for you to learn more about a particular feature, event, place, process, etc., in the world around you that you think is particularly interesting or important. You are dedicating significant time and energies to this project: maintaining the required level of commitments is generally much easier when you are working on something that matters to and genuinely interests you. Having this level of interest and commitment also makes it is easier to communicate your passion for and interest in the subject matter and to write a compelling and engaging dissertation. Being passionate about the subject matter also assists in writing a powerful argument and convincing a reader of the importance and urgency of a topic and that they should invest time in reading about it.

Your dissertation is the capstone piece of work for your degree. As such, it carries a lot of weight for your final degree outcome. The mark awarded to your dissertation can have a significant impact on your overall degree classification (in my department, the dissertation is worth roughly 20% of the final degree grade). On an instrumental level, therefore, your dissertation matters in the sense of your degree outcome. It is in your own interest to ensure your work clearly and effectively meets the requirements set out in your department's marking criteria. However, this instrumental motivation is not the narrative to present in your dissertation.

What you need to communicate to your reader instead is why your work matters on a substantive – be this empirical, conceptual, and/or theoretical – level. Communicating this returns us to the idea of 'standing on the shoulders of giants'. Your dissertation matters because it adds to understanding and knowledge – even if in the most incremental manner. It is key that you know and articulate what this contribution is to the reader.

Your dissertation is also an opportunity to express your creativity and showcase various abilities and transferable skills – which leads to another reason your dissertation matters: employers. Alongside developing and demonstrating your knowledge and research skills, you are also advancing and evidencing a broader array of other, transferable skills. These skills and experience matter: in a competitive job market, employers often view dissertations as a key signifier of your ability to ability to synthesise and communicate information, your aptitude for working independently, and as a general indicator of overall employability (Nicholson et al., 2010). These skills cover project planning and time management, written and verbal communication skills, numeracy and literacy skills, as well as aptitudes in digital capabilities, interpersonal skills, personal development, and more. You may not realise you are developing or using these skills – but they are highly sought-after by employers. It is important then to be able to recognise and communicate your proficiency in these.

In application letters and during interviews you can use your dissertation – the whole process, not just the written document – as evidence of your ability to identify a problem, design a reliable approach to investigate and address this issue, collect and/or analyse relevant (quantitative and/or qualitative) data, and make and communicate an informed set of recommendations and conclusions from this. In other words, your dissertation not only requires academic and research skills but also multiple 'soft' or transferable skills, from critical thinking to problem-solving to time management to communication skills. Further discussion on how your dissertation can be used to demonstrate these capabilities to employers is covered in Chapter 15 – *What next? Employability and your dissertation.* At the same time, however, it is crucial not to lose sight of the primary audience for your dissertation – your examiners – and to ensure that your writing is tailored towards their expectations and requirements (and, by extension, the requirements set out in the relevant marking criteria).

ENGAGING YOUR AUDIENCE

As with any assessment, engaging your audience is crucial. You need to ensure your dissertation is clear and well written (see Chapter 3 – *Writing your dissertation*, and Chapter 4 – *Good writing*) and gets your reader's attention. You want to give your reader a reason to care about your work – draw them in by highlighting why your work matters, why you find the topic important or interesting, and/or explaining how your work advances existing knowledge. It would be oversimplifying things to say your dissertation needs a sales-pitch but it is important that you make it as straightforward and enjoyable as possible for your primary audience – your examiners – to engage with your work (also Chapter 14 – *Remember your audience: Keeping your marker onside*).

A crucial starting point to this process is for you to understand and communicate why you are exploring this specific topic or question. Why does this matter to you? And, more broadly, why does (or should) this matter to others? This may feel intuitive, but being able to explain to your reader what your motivation is or why this particular topic or question is worth their investment of time and energies to read is key. To help communicate this rationale to your audience, it can be helpful to write down your responses to the following questions. You can use these as a reference point during the writing-up process, and may be useful in developing an opening hook or rationale for your introduction:

1 Why did you choose this topic? Why is it interesting and important? Was this choice inspired by existing academic theory or ideas? From a personal encounter or experience? From recent media coverage?
2 Is there a reason this is particularly timely or urgent to explore?
3 How does your work engage with existing academic literature? What is the gap, niche, or silence you are speaking to? Is it a gap in theory or conceptual work, the chance to apply an

existing idea to a new case study, a gap in empirical knowledge, the opportunity to use a new or different method to explore a question or dataset?

Using reflective notes from your answers to these questions can help with the initial planning and structuring of your dissertation, including identifying the core argument or narrative of your work – the 'red thread'. These notes can also help you to explain how your project is timely and important, and to outline why you have decided on this topic – not in a first-person narration of "I decided to do this project because" but in a more abstract sense of "In light of recent news coverage of these concerns, it is imperative to develop detailed understandings of". Understanding the underlying reason for your research may also provide a steer as to where to position your research questions within your dissertation as well as providing potential sources of 'hooks' for the reader.

HOOKING YOUR READER

'Hooks' are narrative devices used to grab the reader's attention. While you may use these various places in your dissertation, most commonly these are seen in the opening of a chapter – the introduction in particular. Hooks may be in the form of an epigraph or quote that pithily highlights a key idea from existing literature, a key quote from a news headline or an interview that provides a provocative or powerful entry point, a short narrative recounting an experience or event, or a rhetorical question that invites the reader to engage with the subject matter. For instance, the opening hook for a dissertation addressing community-led responses to flood risk might

- use the 17 December 2024 headline in *The Guardian* that stated, "More than 6m homes at risk of flooding in England, says Environment Agency" (Horton, 2024) as an epigraph or opening quote;
- use an extract from a research interview with someone whose house was flooded to personalise and contextualise the issue and the importance of addressing it;

- provide a first-person description of your own experience of flooding or the aftermath thereof and the impacts of this;
- ask a rhetorical question such as the following:

> With government funding for flood defences under growing pressure and insurance companies increasingly reluctant to provide cover for home that have previously flooded, what hope is there for the owners and occupiers of the more than 6 million UK homes at risk of flooding?

Any of these narrative devices should help your reader to connect with your project and understand why it is important – they also give you the chance to signpost the focus, structure, and argument of your dissertation.

The use of such mechanisms to gain a reader's attention and interest is not confined to undergraduate dissertations. These are commonly seen in books and articles, as well as in Masters' level dissertations and PhD theses. It is important, though, not to simply use provocative quotes without linking these to the following text. You need to make it clear why you are using these. The following vignette is taken from the opening page of my own PhD thesis (Hammett, 2007) on race and identity in South Africa – on its own, the quote can feel gratuitous or sensationalist.

> While I was at Hewat Training College. We were in what is now Klipfontein Road. We were at the bottom of Klipfontein marching towards Rondebosch, which was a 'white' area. So what the cops did, they covered the 'white' area to keep us out. Now the tactic of the police was that they would shoot the main person in that crowd. They would identify the leader… a sharpshooter would kill that guy and everyone else would disperse.
>
> On this particular day, one of my colleagues… he was shot right in front of me; blood spattered on my face… That evening, we went to this youngster's home. His father was there, and we spoke to him and the father was waiting for his son to come home. The son never came home. It was the same

> sergeant who had given the order to shoot. He had killed his own son. He committed suicide, I think, a month or two later (David, interview 13/05/2005).
>
> (Hammett, 2007: 1)

This harrowing tale emerged during an interview with a participant in my PhD project in which we were talking about the intersection of politics and identity in apartheid-era South Africa, and the legacies of this in the early 2000s. I used it in the opening pages of my PhD thesis to encapsulate why the central theme of my PhD was important, to highlight key themes that emerged during the research, and to set up a series of rhetorical questions to frame the rest of the introduction and narrative. Providing a clear and powerful opening is, however, only one component of a good dissertation.

WHAT MAKES A GOOD DISSERTATION?

The components to a good dissertation encompass your initial research design and planning, execution of data collection and analysis, as well as the writing-up of your project. Various books provide excellent guidance on how to deliver the key features of a good dissertation in the research design and delivery phases, including Kimberley Peters's (2017) *Your Human Geography Dissertation: Designing, Doing, Delivering* and Tony Parsons and Peter Knight's (2015) *How To Do Your Dissertation in Geography and Related Disciplines*.

The focus in this book, however, is on the features and aspects of *writing* that contribute to a good dissertation. There are various components to writing well, all of which ultimately contribute to the key goal of clear and compelling communication that ensures your reader not only understands what you are saying but can see why it is important. It is essential that your dissertation has a clear focus which is retained throughout. A common weakness in many dissertations (or at least the first drafts of them) can be the inclusion of too many tangential ideas, debates, and segues that do not clearly speak to or fit the main argument. Stronger dissertations tend to provide a more clearly defined and signposted singular narrative that addresses the aim(s) and question(s) of the research. This central narrative or argument is often termed the 'red thread': the strand of an idea that runs all the way through the dissertation and around which everything else is woven (see Chapter 4 – *Good writing*) to effectively communicate your key findings.

Effective communication in a good dissertation is primarily about telling a simple story elegantly. This does not mean offering a simplistic and overly generalised argument; rather that you focus on conveying one argument or idea with style, flair, and (most crucially) clarity. In other words: your dissertation should seek to do one thing and to do it well. This means offering accessible and precise details on the focus and aims/objectives of your work, how you have delivered on these, and to narrate this in clear and accessible language. In short, do not fall into the trap of equating complexity of language with demonstration of intellect!

Telling your story requires building your dissertation through a series of key blocks. At the highest level, these are the chapters of your dissertation which form the key structural components. These are briefly outlined below and expanded on in subsequent chapters in this book.

Your *introduction* should convey a sense of the big-picture connection and importance of your project and implicitly convey your passion for the subject. In so doing, you draw the reader into your research, providing pointers to what you have done and found, and set out the flow of the rest of your dissertation.

Your *literature chapter(s)* should provide a detailed synthesis of contextual, background, and conceptual/theoretical materials of relevance to your specific focus. Through a critical appraisal of previous research on the topic you then justify your research questions and focus.

The *research methods and design chapter* should clearly narrate your research-plan-in-action, conveying to the reader that your approach was considered, appropriate, robust, and reliable. Ultimately you need this chapter to reassure the reader that they can trust your data and findings. This chapter should cover your overarching approach to the research as well as the methods of data collection and analysis used, alongside consideration of ethics, positionality, and similar concerns.

Your *results and discussions chapter(s)* should communicate your data clearly and effectively, providing relevant contextual detail and your analysis and discussion of the data. This part of your dissertation needs to share key data *and* your interpretation of this, using existing literature to help nuance and inform the argument you are making. You need to give 'voice' to the data (i.e., quotes, tables, graphs) and your interpretation of this (i.e., telling the reader what it means).

Finally, your *conclusions* need to succinctly reiterate the main argument you have made based on the evidence contained your dissertation and demonstrate how you have addressed your title and research aims/questions. This is not the time to introduce new data or arguments but rather to reiterate what your contribution to knowledge is.

READING THIS BOOK

How to Write Your Geography Dissertation is dedicated to the writing journey of your dissertation. It provides detailed, in-depth support for each stage of the process along with guidance on thinking through the overarching questions and considerations for geography dissertations. The next chapter provides advice on working with your supervisors and accessing further support for your dissertation. This chapter also addresses the thorny question of (generative) AI and dissertations, as well as responding to various commonly asked questions of supervisors. The following chapter then addresses the preliminary stages of planning and starting the dissertation writing process, recognising that this is a long journey requiring careful time-management in the face of competing demands and deadlines. The subsequent focus on good writing provides detailed practical advice on how to make the best use of your word count and how to convey your ideas and argument as clearly and effectively as possible.

The following chapters then offer step-by-step guidance on writing each substantive chapter of your dissertation. These chapters are ordered to reflect the typical process of writing rather than the reading a dissertation – hence the chapter on writing your introduction comes much later in the book, while advice on delivering your literature review comes first. There are two things to note here. First, while the book's chapter order reflects the typical chapters found in a standard dissertation, this should not preclude you giving consideration to more creative structures if they suit your work. Second, while you might opt to write your dissertation in a strictly sequential order, one chapter at a time, this is not the only (nor necessarily best) approach. Like my approach to writing this book (and pretty much everything else I have ever published), you may well write a section of one chapter and then flit to another chapter and back and forth again and again. In short, dip in and out of the chapters of this book as best supports your writing journey.

The final three *step-by-step* chapters address important components of your dissertation document that are not standard chapters – namely the front matter, reference list, and appendices. While these will only be finalised at the end of the writing journey, you are likely to be populating sections of them throughout the process. You should draw on these chapters throughout the writing process – not least in relation to the guidance on referencing which is most helpful if embedded from the start of your writing journey.

The closing sections of the book then reflect on the importance of understanding how your work will be assessed and how to keep your reader (your marker) on-side. Recognising that your dissertation does not necessarily end with its submission for assessment, the subsequent chapter considers how you might consider communicating your findings to other audiences and how to use these materials as a key resource in the job market. The final chapter then provides some closing thoughts and reminders intended to support and (hopefully) motivate your writing.

CHAPTER SUMMARY

This opening chapter has set out to provide a brief overview of why your dissertation matters on multiple levels, and to use this understanding to begin to ask useful questions intended to help structure and narrate aspects of this work. In trailing different ideas and concerns – from 'hooks' and 'red threads' to thinking about skills and audiences – this chapter has set out a holistic approach to thinking about writing up your dissertation with a view to helping to optimise the direct and indirect benefits and outcomes from this piece of work. The following chapters develop many of these considerations in more depth with the intention of complementing the support provided by your supervisor and other resources within your institution. Before turning to these specific areas of focus, it is sensible to think about how to make effective use of the support available to you and to address some of the more commonly asked questions that do not obviously fit into a specific chapter within your dissertation.

2

SUPERVISORS, SUPPORT, AND COMMON QUESTIONS

There are numerous resources available to you to support you during your dissertation journey – this book included. The most important resource, however, will be your supervisor. This chapter offers guidance on how to get the most from your working relationship with your supervisor, as well as suggesting other likely sources of support within your university and addressing various generic questions that are regularly asked of supervisors.

GETTING THE MOST FROM YOUR SUPERVISOR

Your supervisor will be a vital source of support and guidance for your dissertation. However, the time they have to support you with this will be limited. Many departments have rules and expectations around how many meetings you can have with your supervisor, and whether your supervisor is able to read and comment on any or all of your dissertation. Be sure to familiarise yourself with this information, and don't be surprised when supervisors stick to these rules. Ultimately it is down to you to ensure you make effective use of the time your supervisor is able to offer you.

RESPECT YOUR SUPERVISOR'S TIME

Your supervisor will have limited time to spend on your dissertation, so you ensure you respect this. One key aspect to this is to be prepared for meetings. Think about the substantive questions you want to ask, the progress you've made, and the support you need. In other

DOI: 10.4324/9781003584315-2

words, ahead of time think about what *you* want from the meeting: do you want help with consolidating your focus or questions? With identifying or weighing-up the merits of different literatures? With discussing how to write-up a particular section? Use this understanding to create an agenda or at least a set of questions or prompts you want to discuss. It may be that your supervisor is required to cover particular content or materials in a specific supervision, so be aware of and accommodate this – but always arrive with a sense of purpose and ownership for your supervision meeting. Linked to this, don't arrive for a supervision and say 'I've not really done anything on my dissertation since we last met'. Take ownership and responsibility of the journey and carve out some time before the meeting to do something – anything – towards your dissertation and arrive at the supervision with a specific outline or plan or question to discuss. It is much easier for supervisors to give focussed and effective feedback on specific questions and ideas than to offer generalised responses to un- or under-developed abstract questions.

Linked to this, if you are asked to send materials to your supervisor in advance of a meeting then do so – and in good time (i.e., not 5 minutes before the meeting). This gives your supervisor time to read and think, resulting in more in-depth and useful feedback. If you don't send materials in advance or ask supervisors to read them in a meeting this is a poor use of your time and theirs, so help yourself by helping your supervisor. It is also helpful for your supervisor – as well as being simply polite and professional – to let them know if you are going to be late to or miss a meeting or deadline. Staff will have set aside time in their diaries for these meetings or to read materials, so if you cannot make use of this then let them know in good time. This is a simple matter of respect for your supervisor – their time is valuable, and if you communicate clearly to minimise any wasted time they may be more inclined to make alternative arrangements to support you.

HAVE CLEAR LINES OF COMMUNICATION

Be clear in your questions and requests to your supervisors but also be open to constructive criticism and feedback. One of your supervisor's key roles is to provide critical feedback to strengthen and improve your dissertation. A good supervisor will highlight areas of strength and offer positive feedback, but they also need to highlight sections and work that could be improved. While critical feedback is to be expected, this does not mean you need to passively accept all the advice you are given or make every suggested change – if you strongly disagree with the feedback provided then ask for clarification on this and discuss in more detail as appropriate. This is often best done during supervision meetings to allow for a constructive discussion in real-time rather than a disjointed exchange of (complicated) emails.

While email can be a tremendously effective and efficient means of communication, it is not always the best option. Short, simple questions and queries can be effectively asked and resolved by email – *but* always check module handbook and other materials for the answers to generic questions first. Supervisors get incredibly frustrated if you are asking questions to which you have already been given an answer – and you really don't want to irritate your supervisor by doing this. Lengthier, more complex, or discursive questions are usually far more effectively discussed verbally in a supervision meeting or office hour appointment. Long and complicated emails are a time-drain – taking a long time to type and often being difficult to respond to in a simple and succinct manner. You may well find that in response to a long and complex email your supervisor simply responds saying 'book to see me during consultation hours' or 'we will need to discuss in your next supervision'. This is not their attempting to avoid work but instead a recognition that a conversation is needed to properly understand and accurately respond to your query.

Even with shorter emails, do not expect or demand an immediate response from your supervisor: they will be juggling multiple other tasks (teaching, marking, administrative tasks, research activities, etc.), many of which will be time-sensitive and have specific deadlines or meeting times. Recognising this, many departments now have a communication policy that sets out timeframes for responses to emails (e.g., three days). It is always important to check if such a policy exists and what it states. In general, however, if you have not had a response after three working days, it is acceptable to send a polite reminder/follow-up message. (It is also worth remembering that an email cannot easily convey tone and intent – at times this can lead to emails being interpreted as overly defensive, aggressive, accusatory, or demanding. It is always worth pausing to check and – as necessary – edit an email to try and avoid unintended offence being taken.)

TAKE RESPONSIBILITY FOR YOUR DISSERTATION

Supervisors cannot give feedback on ideas or work you haven't done. It is not their job to design or write your dissertation for you. It is easier for supervisors to offer substantive feedback on things that are written and/or to specific questions. For instance, rather than asking your supervisor 'How should I structure my discussion?', develop your own initial plan or outline structure and use this as a prompt for detailed and specific discussion. Linked to this, try to avoid questions such as 'How do I write X?' – to which the facetious answer many supervisors want to give is 'by putting words on paper'. Instead develop a plan (or plans) for how you might structure and write the section and ask for specific feedback and comment on this.

TAKING NOTE(S)

It is amazing how often students will attend a dissertation supervision and never write down or record any of the conversation or feedback provided. No matter how good your memory is, it is highly unlikely that you will remember all the details, ideas, and other input provided by your supervisor a few hours or days later. It is important then to make notes or – with the permission of your supervision – to record supervision meetings to help ensure you don't miss key feedback or useful prompts. You should also set aside time as soon as possible after the meeting to expand and refine these notes, develop action points and timelines, follow-up on lines of thinking, and so on – and if you realise you aren't sure about something, the sooner you contact your supervisor to get clarification the more likely they are to be able to recall the conversation and ideas discussed.

Supervisors are not there to do the following:

- Design or write your dissertation.
- Be your friend. Our role is to offer constructively critical feedback and support your dissertation.
- Proofread your work and correct spelling, grammar, punctuation, or typographical errors.
- Answer questions that are already answered in the module handbook, marking guidelines, or other materials provided to you.
- Respond immediately to your emailed questions or requests.

ADDITIONAL SUPPORT

While your supervisor is your main source of academic guidance, they are not the only support available to you. Your university library will likely offer various kinds of support – either through online training resources, support sessions, or potentially even direct one-to-one assistance. This support can encompass guidance on effective literature searching, archival access, access to and training in specific bibliography/reference management software, effective data management strategies, critical thinking, or writing (and reading) skills support. Take time to check what support is available and make use of these resources where helpful.

Further support and resources may be available from other services within your university. One of the most under-rated sources of support is English language support. While commonly perceived as being a resource for students for whom English is not their first language, many language support centres offer training and support to all students – either through group (training) sessions or bespoke, individual consultations. There is no shame nor stigma in seeking expert guidance and support with writing and it can be transformative for your grade. The most useful training I undertook as a PhD student was a one-day effective writing course with the English language support centre at my institution. What I learnt during that course helped me write my PhD thesis more effectively and continues to influence how I write now. Your institution may also have support services to assist with other academic skills, statistical methods and analysis, training in appropriate use of AI, and so on. In short – it is worth spending a bit of time scoping out the support offered by your university and identifying if any of this would be beneficial for your dissertation (and studies more generally).

If your department allows this, you may also find useful to seek additional, focussed input from academic staff other than your supervisor. Expectations on this can vary significantly between departments – in some departments this is explicitly prohibited, in others it may be tolerated or even encouraged. If it is accepted practice in your department, then any approaches to non-supervisory staff should be occasional and focussed on a discrete and specific question or request for specialist support (for instance, methodological or analytical expertise, insight into a particular debate). But again, be respectful of their workload – these requests are exceptional and need to be occasional and specific. Increasingly, students are also using artificial intelligence (AI) as a further resource to support their dissertation work.

> Other support mechanisms can be more informal, including peer-to-peer support and asking family and friends to read a draft of your dissertation and respond to respond to specific questions such as:
>
> What was the main argument of the dissertation in two sentences?
> Does the title match the content and conclusion?
> Does the introduction accurately and clearly set-up the rest of the dissertation?
> Was there anything that was unclear? Were any sections overly repetitive? Were there any sections that felt tangential or irrelevant to the main argument?

AI AND YOUR DISSERTATION

There is no escaping from the reality that artificial intelligence (AI) is playing an increasing role not only in society but also in (higher) education. Staff and students have long used Google Scholar and similar platforms in searching for literature – while not AI tools themselves, these do use AI to improve search results. Likewise, autocorrect in word processing programmes or tools such as Grammarly are well-established and commonly accepted forms of predictive AI. More recent developments – including AI-based transcription tools and generative AI (GenAI) – have raised more critical questions and challenges for students, academics, and institutions. While the case can be made to prohibit any use of various forms of AI in academic work, many universities are adopting a more pragmatic or critically progressive stance, recognising AI will play an increasing role in society, work, and education in the coming years. Resultant institutional policies outline permitted or 'fair' use of AI.

The starting point for any discussion of the use of AI in your dissertation is to check what your institutional guidance is and whether there is further guidance or instruction on the use of AI within your department or for your dissertation specifically. While there is variation between universities in specific policies, in 2023, the Russell Group of universities agreed a common set of underlying principles on the use of GenAI:

- Universities will support students and staff to become AI-literate.
- Staff should be equipped to support students to use generative AI tools effectively and appropriately in their learning experience.
- Universities will adapt teaching and assessment to incorporate the ethical use of generative AI and support equal access.
- Universities will ensure academic rigour and integrity is upheld.
- Universities will work collaboratively to share best practice as the technology and its application in education evolves.

(Russell Group, 2023)

Institution-level guidance alongside these principles often emphasises the following:

- A requirement that you record all prompts used and on which GenAI platforms as well as the generated response. This is a key evidence trail in case of any allegations of misuse of GenAI – so it is in your own interest to do this.

- The need to critically engage with the materials produced by GenAI in response to prompts, questions, and requests to develop ideas. You should never simply cut-and-paste responses into an assignment (unless specifically told you can do).
- Only use permitted AI platforms (for instance, at the University of Sheffield this is Google Gemini, at King's College London, it is Microsoft CoPilot) and using recommended settings.
- The requirement to fully acknowledge and declare any use made of AI in your work.
- The expectation that if you are unsure of how to or how you are permitted to use AI you should always check with relevant staff for guidance and confirmation. It is *your* responsibility to ensure you are following the rules.

If you are considering the use of GenAI to support you through your dissertation journey, it is important to do so critically and to be aware of various limitations and implications. The use of GenAI to produce summaries of debates or definitions undermines the development of your own research and critical thinking skills. These are valuable transferable skills that you need to develop – an over reliance on GenAI will ultimately risk being a limiting factor for you in the workplace. Of more immediate concern, however, is the risk of AI-generated summaries being inaccurate, flawed, and unreliable, and/or that they may perpetuate particular (dominant) voices and provide a skewed overview on the topic. Without your own detailed and critical engagement with the literature and arguments, you cannot be sure of the veracity of the generated material, nor will you be able to as effectively utilise theoretical understanding in your analysis and discussion. You therefore need to critically engage with and reflect on any GenAI materials and consider whose voices are being included or amplified, whose are being neglected, but also to consider what *your* understanding of a topic is and what the argument is that *you* want to make.

Critical interrogation of AI-generated content is also required to mitigate against the risks of 'hallucinations' and invented citations and references. There have been numerous scandals in recent years of papers being rejected by or withdrawn from academic journals for misuse of GenAI, and court cases in multiple countries have been blighted by false AI-generated cases and precedents. You do not want your dissertation to be flagged for misuse of AI and academic misconduct!

Do not use AI to run analysis of your data or to write your conclusions.

Not only can the outputs be flawed and illogical, but they are unlikely to link to your framing literature. There is also the risk that the AI platform may create and add-in fictitious data. There is also a significant potential that providing confidential or anonymised data to an AI platform may breach research ethics and data management approvals. If this occurs you are likely to face severe academic misconduct outcomes

It is also important to be aware of and consider the environmental impact of using GenAI. These impacts include indirect impacts arising from increased demand for high-performance computing equipment and the training of new GenAI models as well as the direct 'costs' from each query or request. While individual requests may have a limited carbon cost (generating one AI image uses the equivalent energy as fully charging a mobile phone; 1,000 text generation requests use the equivalent of 16% of a phone's charge; Heikkilä, 2023) the cumulative costs of training and running GenAI is vast: it is estimated that "by 2035, the global carbon emissions attributable to G[enerative]AI will range from 18.21 to 245.94 million tons" (Ding et al., 2025: 2667). Layered on to the energy and carbon costs (sustainable energy sources cannot meet the current rate of demand increase linked to GenAI [Zewe, 2025]) is the impact on water supplies as 2 litres of water are required for cooling purposes for each kilowatt hour of energy used. As water stress increases, these demands place an increasingly challenging burden upon water supplies and ecosystems.

If you do opt to use any form of GenAI in your dissertation then be incredibly careful to adhere to your institutions polices (including permitted platforms) and use it in very specific ways with clear prompts, and to treat generated materials critically. You need to keep a record of these prompts and the generated outputs – not simply because this can be demanded by your institution, but also so you have an evidence trail if you are asked to account for your use of GenAI. And remember to fully acknowledge and declare any and all use of GenAI.

AI-TRANSCRIPTION TOOLS

In addition to the use of GenAI, another growing area of AI use relates to transcription. The process of transcribing interviews and similar methods is incredibly time-consuming: it typically takes 6–8 hours to transcribe a single recorded hour of an interview. The rise of AI-based transcription tools has the potential to be a game-changer in providing significant time savings. However, there is a catch. Content uploaded to an AI-transcription tool may then be used in the large language machine learning process and beyond. This presents clear risks in relation to research ethics, data management, privacy, confidentiality, and so on. Consequently, many institutions have policies on which AI-transcription software may be permitted by staff and students. As a result, you may find that common platforms, such as Otter.ai, Trint, Turboscribe, and Verbit, are not permitted for use by your institution. Instead, you may need to use a platform within your institution's suite of IT resources. This may be a bespoke transcription tool where you can directly upload recordings for transcription. Or it might be through using Microsoft Teams, Google Meet, or equivalent to host a 'meeting' and request a transcript of the meeting and then play your audio recording into the meeting space. This is a more time-consuming approach but is nonetheless significantly time-saving compared to traditional transcription approaches.

If you do use AI transcription, you must do the following:

1. Always read and cross-check the transcript against the original recording. These platforms rely on predictive text and will make mistakes, particularly when dealing with recordings that are heavily accented or that use slang or colloquialisms, etc. It is your responsibility to check and correct.
2. Acknowledge that you have used AI transcription – including detailing which platform(s) were used.
3. Remember – if you use non-approved software you risk breaching of your research ethics approval and may face academic sanctions, so be sure to use an approved option.

FREQUENTLY ASKED QUESTIONS

Supervisors regularly get asked similar questions each year. Some relate to specific dissertation chapters (these are addressed in the relevant chapter(s) in this book), while others are more general. Some of these questions are addressed here.

DO I NEED TO HAVE (A) RESEARCH QUESTION(S)?

Yes (and no). You need to have a clear research aim and specific objectives. These are often most easily communicated in the form of question(s) which allow the marker to understand what you are trying to find out and whether you have succeeded. Crucially, check the relevant marking guidelines in your department – if they specifically mention 'research questions' then be sure to use this language. Not all departments use the terminology of 'questions' in their criteria, others talk of 'outcomes', 'opportunities', or 'project aims'. In such a situation you might think about whether using the terminology of research questions is best suited to your work, or whether a different phrasing of aims or objectives might be more apt. Either way, you must have a clear focus and aim that frames your dissertation and provides a sense of coherence to the project.

WHICH SECTION SHOULD I WRITE FIRST?

As intimated in the previous chapter, you do not write a dissertation in the same sequence as you would read it. As your dissertation proposal and planning will have been grounded in existing literature, this is often a useful chapter to develop early in the writing process. You should already have plenty of notes and an understanding of the key debates, so turning this into a chapter can generally be done early. Developing this content will in turn help develop and inform your analysis and discussion – another reason why drafting your literature review chapter should be done sooner rather than later.

Your research design and methods chapter is another common option to draft first – with much of the content here a description of what you have done, this can be a relatively straightforward chapter to work on (but see Chapter 6 – *Step by step: Research design and methods* – on the importance of going beyond simply describing or recounting what you did) and get into the writing phase with. Drafting this content soon after completing fieldwork and/or analysis also helps ensure you are able to accurately recall what you did and reflect on challenges and limitations more readily.

It is likely you would then turn to the results and discussion chapter(s), drawing on your analysis and literature to inform and nuance your discussion. Following this, the final chapters to draft will be your conclusion and introduction as these can only be fully written once the other content is drafted and your argument solidified.

WHERE SHOULD I PUT MY RESEARCH QUESTIONS (OR AIMS/OBJECTIVES)?

These need to be made explicit and clear in your dissertation – but exactly where to include them is a matter of debate and subdisciplinary preference. There are four common locations for research questions (or aims or objectives or hypotheses – for the sake of simplicity, I will use research questions as an overarching proxy from hereon) in your dissertation: the introduction, a short standalone 'aims' or 'questions' chapter, the literature review, or the methods chapter. There are pros and cons to each option, and your decision will be informed by consideration of flow, logic, supervisor/departmental direction, and the origins of your questions.

Introduction

Including your research questions in the introduction can help frame the focus and structure of your dissertation from the outset. In this instance, you would typically locate your research questions towards the end of the introduction, but before you outline the structure of the dissertation and substantive focus of each chapter. This provides a frame of reference for your reader's engagement with the subsequent literature review and through the rest of your dissertation. You can then explicitly use the research questions to justify why you are engaging in depth with specific literatures, why you might only be signposting other materials and debates, and why you are developing your thesis in a particular way.

Potential disadvantages of this option include the risk that your research questions may appear to lack foundations in or connection to existing literature and debates. To counter this you can add a succinct opening framing for the dissertation that strategically signposts key debates that you return to in the literature review. Placing the research questions this early in your dissertation can also mean that the marker has forgotten the specific details of your questions by the time they reach the data and discussion sections. You may then find it necessary to repeat or at least signpost back to them later in your document.

Research aims 'chapter'

Some dissertations include a very short chapter after either the introduction or the literature review that is specifically titled and solely focussed on the research aims, objectives, questions, and/or hypotheses. When this approach is used, the chapter typically consists of a brief outline of the overall aim of the dissertation (providing a succinct rationale for the focus and connection to existing debates) and a list of research questions/hypotheses/objectives. This approach offers the benefits of a clear identification in the contents page of where the aims/questions are located and to effectively bridge between other chapters. To be effective, however, care is needed to craft a clear opening paragraph to introduce the questions and rationale for them, and to signpost how these then link to or inform the following chapter.

On the flip side, including aims and questions in a discrete chapter can lead to these feeling isolated, disjointed, and incomplete if the chapter is simply a list of aims and/or questions. As discussed in more detail in Chapter 10 (*Step by step: Introduction*), wherever you include your research questions, it is important to provide some framing and linking content.

Literature review

Outlining your research questions at the end of the literature review can be effective if your questions are based on and derive from the identification of a gap within existing literature. This approach can allow you to build an argument and narrative from your critical reading of the literature which then underpins your (choice of) research questions. Presenting your research questions here can also provide an effective segue to your research design and methods, ensuring the reader has this information fresh in their mind.

However, if your research questions are not primarily driven by or informed by the literature, they can feel very out of place if included here. There is also a risk that without a careful curation of the literature review and signposting to your research questions that your marker may not have a clear sense of why you are discussing a particular set of ideas and debates. To counter this you need to ensure a clear narrative throughout the literature review that builds towards your research questions.

Methods chapter

Placing your research questions at the start of the methods chapter can allow you to connect back to the literature chapter while linking to your methods and rationale for these. This can be effective option if the questions don't immediately link to your introduction or literature. However, leaving the questions until the methods chapter can risk the opening sections of your dissertation feeling ill-focussed or lacking in purpose. As with the advice relating to locating your research questions at the end of the literature review chapter, you would need to very clearly frame and signpost the flow and focus of the opening chapters of your dissertation to give this sense of clarity and purpose.

Presenting your research questions

Regardless of where you include your research questions you must ensure they are clear and concise, and that your dissertation addresses them. Do not just drop in a sub-heading that says 'Research questions' and list your questions with no linking sentences to the rest of the chapter content. Adding a couple of sentences that make the connection with surrounding content, including justifying your questions in relation to existing debates, can make a substantive difference to your work. For more guidance on writing and presenting research questions – regardless of where they are located in your dissertation – see the relevant sub-section in Chapter 10 – *Step by step: Introduction*.

Wherever you include your questions there are some common considerations in how you present them. First, use a clear (sub)heading to identify where you are including your research aims/questions and include this in your detailed list of contents (see Chapter 11 – *Step by step: Front matter*). Second, set out a clear overarching aim or rationale of what you dissertation is doing and how (either in terms of methods or topics). This should take no more than five sentences. In other words, it is a really abbreviated introduction to and rationale for your questions.

Your research questions should follow on and be clear and focussed, contribute to your overall aim, and be answered within your dissertation. You will have started your dissertation with a set of research questions to address – but these may have evolved and developed during the research process. It is really important to ensure that the questions you include here are the ones you have actually addressed in your dissertation.

Typically, you might have between two and five research questions for a dissertation, each addressing a constituent part of the overall research aim. These may be further developed with more specificity through a series of secondary research questions or hypotheses. These sub-questions should each provide a discrete building block of understanding towards your overall argument.

When writing your research questions, you should ensure the following:

- They are answered within your dissertation. Ask yourself whether they are clearly defined in scope (be this geographical, temporal, scale, concept, etc.) and check that your findings have answered the questions outlined.
- They are clear and singular. Is it obvious what the question is focussed on? Have you avoided using sub-clauses or multiple foci in the question? If not, think about splitting the question into more specific parts – potentially as sub-questions.
- They are not all closed questions leading to 'yes or no' answers. Think about the wording of your questions: in some instance, a 'Does' question is appropriate, in other situations, 'To what extent does', 'Why does', or 'How does' question may provide greater scope for analysis and critical explanation.
- Each question is relevant to and helps build a response to the overall research aim/focus.

When presenting your questions or hypotheses it is important to clearly differentiate between primary or secondary questions. This can be done through simple formatting techniques as well as through the use of a hierarchical numbering strategy:

> **Example 1:** This research aims to understand the impact of soft-engineering approaches to riverbank management on macroinvertebrate diversity in the River Don, Sheffield. To realise this aim, this work addresses the following hypothesis:
>
> **Hypothesis 1:** Soft-engineering approaches to riverbank management results in an increase in macroinvertebrate species abundance and diversity in the River Don.

> **Example 2:** This research explores how geographies of fear affect students' use of public spaces in Huddersfield. This aim is addressed through the following research questions:
>
> **RQ1:** What are students mental maps of fear and safety of Huddersfield town centre?
>
> - RQ1.1: Which areas of Huddersfield town centre are viewed as (un)safe?
> - RQ1.2: How do these mental maps differ among groups by protected characteristics such as gender, sexuality, religion, racial identity?
>
> **RQ2:** How do these understandings of Huddersfield's urban geography affect students' use and navigation of these spaces?
>
> - RQ2.1: What practical measures do students take to navigate around or through these spaces?
> - RQ2.2: To what extent does time of day affect these strategies and practices?
> - RQ2.3: What measures could be put in place to improve students' sense of safety within and use of public spaces within Huddersfield town centre?

However you structure the presentation of your research questions, and wherever they are located within your dissertation, it is crucial that they are integral to the red thread of your writing and are answered through the data, discussion, and analysis presented.

HOW LONG SHOULD EACH CHAPTER BE?

There is no definitive answer to this question as it depends on the style and type of dissertation and the relative strengths and weaknesses of your work. The majority of your dissertation word count should be dedicated to your literature and results/discussion and analysis sections. Your methods chapter needs adequate space to allow you to demonstrate that your research is robust and reliable but is unlikely to be the primary focus or contribution of your dissertation.

As a very rough starting point, and assuming the front matter, reference list, and appendices are excluded from the word limit, a useful guide would be as follows:

- Introduction – 5–10% overall word limit
- Literature section(s) – 30–40%*
- Methods/research design – 15–20%
- Results and discussion – 30–40%*
- Conclusion – 5–10%

This guidance should *not* be taken as *the* correct approach but rather as a starting point for writing: the balance of content between chapters will vary depending on where specific content fits best and the relative weighting in your work towards theory or empirical materials.

* For quantitative and/or laboratory-based dissertations, the literature review may be somewhat shorter, with extra space provided for detailed laboratory or other processes and/or to support longer results and discussion sections.

HOW MANY REFERENCES SHOULD I HAVE IN MY DISSERTATION?

The perennial question that is inevitably met with the 'how long is a piece of string?' response. There is no hard-and-fast rule on this, and much will depend on your sub-disciplinary area and style of dissertation. More important than the absolute number of cited works, is how well you make use of these – and this may be in multiple ways. Some citations may be used effectively to signpost wider debates or tangential fields of study – and that is fine. Other citations will be linked to more substantive discussion and analysis.

Both types of citations are important – but it is vital to find an appropriate balance as part of an overall aim of developing a nuanced, detailed, and critically informed engagement with and argument from existing literature.

IS IT A REFERENCE LIST OR A BIBLIOGRAPHY AT THE END OF MY DISSERTATION?

Assuming it is a list of the sources you have cited it is a reference list. A bibliography is something different – it is a list of sources on a particular topic (not just the ones you have cited), and this is not what is needed in a dissertation.

WHAT DO I DO IF THERE IS A SUDDEN CHANGE IN THE CONTEXT OF MY RESEARCH?

Firstly, don't panic – this isn't a problem. You set your research questions and collected your data based upon a particular context and set of conditions. If things have unexpectedly changed or new policies have come into play, your supervisor and markers will understand: you cannot be expected to predict the future. The simplest approach is to add a short disclaimer or note to you introduction that acknowledges the significant change but this occurred after your project was designed and substantively conducted, and that your analysis is still valid in providing a critical engagement with the previous context. For example, you might be working on a dissertation on the impacts of unconditional cash transfer policies on poverty levels in Zambia, but two months before you are due to submit your dissertation a key policy is updated or replaced. This happened to one of my PhD students – and we provided this same advice to them: the policy change does not mean your work is invalid or redundant. Instead, briefly acknowledge that the policy framework has changed but that this occurred too late within the project to be addressed. If you want to take the point further you can suggest that this policy change offers a new opportunity for further research (beyond the scope of your dissertation) to develop a comparative understanding of reactions to or outcomes of the policy change.

IN CLOSING

This chapter has offered guidance on how to effectively manage your working relationship with your supervisor, providing suggestions on how to get the most from their input and the benefits of this. Beyond your immediate supervisor, there are many other forms of support available to you and this chapter signposts some of the more common resources you might wish to access as well as reflections on the growing prevalence of (generative) AI and its implications for your dissertation. Following from this, the chapter offers responses to various of the generic questions that are among the more commonly asked of supervisors. The next chapter turns to considering how to begin actually writing your dissertation.

3

WRITING YOUR DISSERTATION

As noted in the previous chapter, one of the more common questions asked of supervisors is a variant on "what should I write first?" While many students begin with drafting their literature review and methods chapters first, there is no singular rule to adhere to. Nor is there a need to work in a purely sequential way, one chapter at a time. You may find it most effective to be working on multiple chapters at once as ideas emerge and links develop between sections. Whatever your approach to writing, a key challenge for many students – particularly with dissertations – is starting the document.

To help overcome this initial barrier, it might be useful to reflect on some of the following questions:

- Do you like to plan from the beginning and work through in order? Do you tend to start from the end – identifying your conclusions and then writing to fit them? Do you like to write with a detailed plan or freestyle?
- How do you develop and plan your ideas? What mind-mapping approaches are effective for you to develop and maintain a clear focus to your writing?
- What medium do you find best for getting initial ideas out? Paper? Computer? Audio recording/dictation?
- Do you prefer to write in the morning or the evening?
- Do you write best when working from detailed reading notes or to 'free write' and then add notes in?

Your answers to these questions will help you identify an effective writing strategy. The rest of this chapter is intended to support this

DOI: 10.4324/9781003584315-3

by providing advice on when, where, and how to start (and finish) writing to help you hand-in your dissertation on time.

MANAGING YOUR TIME AND JUGGLING DEADLINES

Starting with the end may seem counter-intuitive but is highly logical: if you don't submit by your submission deadline you will be penalised or potentially not assessed at all. It is crucial to get your submission deadline into your diary now (if you have not already done so). If you are doing this relatively early in your dissertation process, you may be looking at a deadline many months in the future (longer if you do a year abroad or on placement). Unlike the proposal stage for your dissertation – which likely involved structured contact time and tight deadlines – you may well find the writing-up of your dissertation to have relatively fewer externally imposed deadlines and interim deliverables. This requires you to manage your time more independently and use project management skills to deliver this major piece of writing. If the deadline is several months or more away it can be tempting to think there is no urgency or need to start writing your dissertation yet. Do not give in to this temptation. Once you factor in other commitments you have less time than you think to complete your dissertation. You do not want to end up joining the ranks of dissertation students who have postponed the writing processes only to later offer up the lament of: "I wish I'd started writing earlier – the deadline seemed so far away but now I'm running out of time".

Common contributory factors here are both an underestimation of how long it takes to write well (see Chapter 4 – *Good Writing*) and a failure to recognise the hours, days, or weeks needed to deliver other commitments and assessments. To help pre-empt this challenge, you should plan ahead and map out other time commitments – be these other assessments (and the time needed to prepare these), teaching timetables, paid or voluntary work, holidays, etc. Mapping out these demands will help you visualise the time available to you for your dissertation and help you to develop a schedule of milestones and interim deadlines for your writing. To illustrate this process, Figure 3.1 offers a notional high-level time mapping

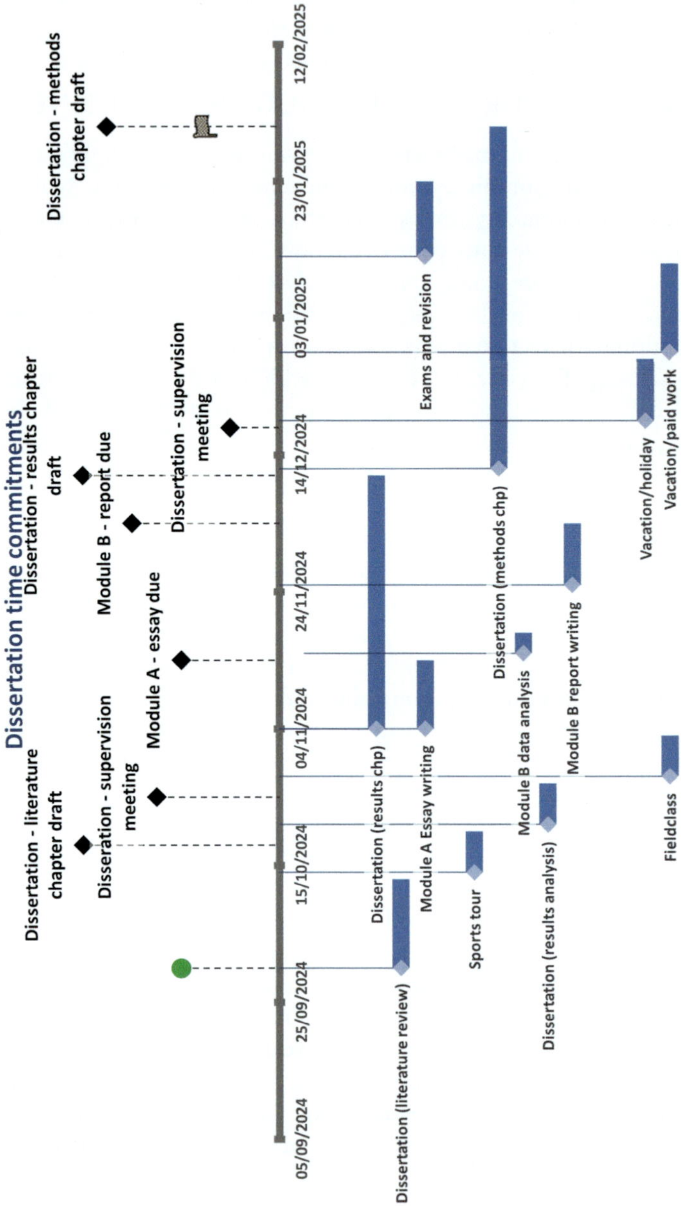

Figure 3.1 Time Management and Milestone Planning for Semester 1 (Assuming Dissertation Submission Late in Semester 2).

and milestone outline for one semester of a dissertation project assuming that data collection has been completed.

Alongside this longer-term strategy, it can be helpful to plan on a day-to-day level, factoring in different commitments and activities on a more granular level – for instance, lectures and seminars and other contact time, and paid or voluntary work shifts. This more granular planning will also then help you identify when to set aside concerted time for writing up your dissertation, while also recognising your writing rhythms (see the following section on *When to write*).

To help you with planning your own write-up, use the following questions and prompts to develop your own timeline to submission:

- When is your submission deadline? Now work backwards and aim for a full-draft to be finished one to two weeks before this deadline to allow time for feedback (from supervisors [if allowed] or family or friends [if possible]) and for final editing and proofreading. This self-imposed deadline also provides a buffer to help with addressing any last-minute crises or illness.
- What other assessments (and assessment deadlines) do you have? Add these deadlines as well as sufficient (be conservative, not optimistic) time to complete the work needed for these assessments.
- What other modules are you taking? What contact time is needed for these? Do any of these involve fieldtrips or other extended periods of working? You may not know the exact number of hours or days needed for these modules but include notional time for these at least.
- Depending on when you are planning the process – have you completed your data collection and analysis? If not, what time is needed for this and when do you intend to complete this?
- What other commitments do you have (sports teams, paid work, volunteering, etc.) and what time do these require? Have you factored in time for rest and relaxation – including holidays?
- Are there interim deadlines you need to meet for your dissertation (e.g., materials to be prepared for supervision meetings, deadlines for a [partial] draft)? If there are no interim deadlines or milestones set of you, then consider setting some of your own (see following section on *Project milestones*).

As you work through this planning, it is important to recognise your working style and how this will affect your other time commitments and the best way to allocate time for your dissertation. For instance, are you a last-minute, pressure worker? Someone who works best in short, intense periods of concentration on a project before needing to switch to another? Or someone who needs a bit of time to get 'into' a project and then spend a longer period of consolidated time on that project? Recognise your preferred working pattern and factor this into your planning and identification of project milestones.

PROJECT MILESTONES AND KEEPING MOMENTUM

Milestones are key targets or deliverables that help you complete a much larger piece of work – they are a common part of project planning across many professions. For your dissertation, milestones might include targets and deliverables during the data collection and analysis processes, such as completing specific data collection activities, finishing transcribing your interviews, or running regression models on your dataset. In the writing phase, milestones might include completed first drafts of individual chapters or sections of chapters, completed second/final drafts of individual chapters, a completed full-draft of your dissertation, and – finally – the submission of your dissertation.

Setting out milestones (as in the example timeline in Figure 3.1) allows you to think about your dissertation in manageable chunks and to allocate time for each discrete deliverable. Having a series of milestones and deadlines allows you to monitor your progress, adapt plans as necessary, and help keep momentum with your dissertation. You may have imposed deadlines from your supervisor to submit various pieces of writing. If not, you should still map out the different sections/chapters of your dissertation that need writing and identify dates by which you want to complete (drafts) of each of these. In doing this, think about how long will each section is likely to take to write and by when do you want to have each draft finished? What about the time needed to receive feedback and complete revisions?

If useful, share these milestones with your supervisor for a further sense of accountability and use these to inform how you prepare for and identify support needed in supervision meetings (see Chapter 2: *Supervisors, support, and common questions*). If you find yourself falling behind your milestones don't panic – use this as a realistic moment to reflect on why this might be happening (perhaps another commitment is taking more time than planned, you had to complete some unexpected extra reading, you spent more time socialising than intended, and so on). From this, consider what your next steps are. This might be to revise your planned timelines, to reconsider the scope of your dissertation, or to adapt other commitments to provide more time for your dissertation work, etc. The sooner you identify challenges in writing up, the greater scope you have to adapt plans and deliver a quality dissertation without a last-minute panicked rush to submit.

CARROT, NOT STICK

You can also use milestones as a positive motivation – allowing yourself (micro)rewards for meeting milestones can be a wonderful motivator. What forms these 'rewards' take are a personal choice – it might be to rent a movie or a trip to the cinema, to enjoy a particular treat (my go-to rewards are jelly beans), to take an afternoon off and go climbing or walking or to feed the ducks in the park – and they should be a positive reward for success and progress. In other words, celebrating and rewarding yourself for successes – whatever that reward looks like for you – can turn deadlines into 'carrots' rather than 'sticks' to help you write.

WHEN TO WRITE

There are two key temporal scales at play in thinking about the question of when to write: the dissertation journey overall and the day-to-day. As noted previously, you should plan your overall dissertation journey and think about how to allocate time for writing and revising each section. As part of this, you might find it useful to consider how you might work on different components of your dissertation in parallel – whether this is in drafting written content while completing data collection or analysis, or in working on multiple chapters concurrently. Some sections of your work can only be written after you have completed other activities – you cannot write up your results until you have completed your data collection and analysis for instance. But you can – and arguably should – be drafting content or at least notes and bullet points for your literature review and methods/research design chapters as you progress with data collection and analysis.

On the day-to-day level it is useful to plan and allocate time for your dissertation – how this plays out on a daily or weekly basis will vary due to other commitments. However, it is useful to recognise that we each have our own daily work rhythms – some imposed (such as work shifts and lectures), and some based on personal preference or natural daily rhythms. Where possible, try to work in sync with your daily and other rhythms (the menstrual cycle, for instance, can affect levels of creativity). Identify what times of day you find creative thinking and writing easier or harder and try to plan your work around these, and you are likely to be more productive and less frustrated with your writing progress. When I was a student, I found mornings (8 am–11 am) and late afternoons/early evenings (4–7 pm) the best times for writing and creative thinking. Early afternoons, by contrast, were (excuse the pun) a write-off: I would struggle to formulate sentences and get frustrated with myself. I learnt that this was the time for me to do something else – to either use this time to make notes on further readings or edit existing text, or to take a break from my studies and climb or run or meet friends for coffee. In simple terms, if possible manage your time to work *with* variations in your concentration and energy rather than against them.

HOW LONG TO WRITE FOR

Continuing with the theme of understanding your work rhythm, it is also helpful to recognise how long you can sustain good writing for. Do you work best in relatively short bursts or longer, more sustained blocs of writing time? Recognising that external constraints may be a limiting factor, try to designate your dissertation writing time in blocks that suit your concentration span and writing style. Sometimes a 5-minute burst of writing might help get an idea on paper, but more commonly you will need a reasonable block of time to get into the flow of writing – but for you, is this an hour, 3 hours, a whole day?

Where possible try to write when, where, and for as long as you feel productive. Crucially, try to write with a degree of regularity or frequency. The longer you leave between periods of writing (in terms of weeks not hours or days) the more time you'll need to get back up to speed and remind yourself of where you are with the process. This can be incredibly time-consuming and frustrating as your writing time is spent re-reading previous content or reminding yourself of previous thoughts and decisions. To combat this, look to designate regular periods of time for writing. Even then, if you are struggling to write then take a break – make a drink, go for a short walk – and get away from the screen or the page for a short while and then come back and try again. Spending hours fruitlessly staring at a blank page or screen will achieve nothing apart from wasting time and a growing sense of frustration and angst. The important consideration here is to ensure you have dedicated time for the writing process, and to accept that how you use this time will vary: it may be to draft some new text, to edit existing writing, to check your referencing, and so on.

The key take-home message here is simple: start writing early and write often. Writing is both a habit and a skill. Getting into the routine of regular writing (and writing with *a* purpose – whether as a stream of consciousness, notes on your methods, an outline of an ethical dilemma, some ad hoc analysis of initial findings, etc.) will be hugely helpful for your dissertation and beyond. It is not only the timing and regularity of your writing that can support your productivity, but also the environment in which you are working.

WHERE SHOULD I WRITE?

Write wherever you can and are most productive. This is not a challenge, but an invitation to think about where best suits your thinking and writing styles. Do you like to be somewhere quiet? Somewhere noisy? Are there times when you want to write on paper or on a computer? Personally, I find it hard to concentrate and write in busy, noisy spaces. However, I have colleagues who find those environments helpful in providing an ambiance and 'energy' that inspires them to write.

Wherever you find best to work, think about both your impacts on that space and the impacts of that space on you. If you like writing in a coffee shop or other public space, think about who else can see your screen or your notes. Are there any concerns with confidentiality you need to take steps to address? But also, be aware of the space you are taking up and be considerate to the business and other customers: buying a double espresso and then occupying one of five tables in a small coffee shop for four hours over the busy lunchtime period is not going be well received.

Other common considerations in managing you working environment include distractions: if you find yourself easily distracted by emails and other notifications, turn these off and go 'offline'. What about background noise? Would using ear plugs, ear defenders, or noise-cancelling headphones to shut out background noise help? Or do you work better with music or other ambient noise? Think about lighting levels – poorly lit or overly bright workspace can cause eye strain and make it harder to concentrate. What about your posture and comfort – are you sitting comfortably or are you risking back, shoulder, or neck pain from being hunched over a laptop? Or wrist pain from typing without any wrist support? These may sound like typical middle-aged concerns, but these types of aches and pains can and do affect many students and can be quickly debilitating and significantly hinder your work. Depending on the space you are using, you will have varying levels of control these factors – but all are useful to be aware of.

WRITING TOGETHER

Your dissertation is an individual, independent piece of work. But it does not need to be written in isolation. Writing together can be a helpful approach, providing social contact and peer support, as well as a sense of accountability and routine. Consider organising regular (weekly) informal 'writing retreats' with a few course-mates for a few hours at a time to collectively write your dissertations. You might find it useful to open these sessions with everyone sharing their writing aims for the session before spending time writing, with the final few minutes spent sharing and reflecting on progress, concerns, questions, and so on.

PLANNING YOUR WRITING

The previous sections have approached planning your writing journey in terms of timelines and deadlines. The focus now turns to planning your dissertation and advice on how to plan your dissertation itself as a document. Initially, this may seem intimidating but using the adage of 'How do you eat an elephant? One bite at a time', we can think about planning your dissertation through a five-step (bite?) process:

Step 1 – Outline the key materials you need to include in your dissertation – from key theory and seminal sources to empirical materials, from methods to themes from your analysis.
Step 2 – Identify your key chapters.
Step 3 – Map your initial content areas (from step 1) onto your chapter structure (from step 2) and assign an estimated word count to each chapter based on (i) overall word limit and (ii) material that will go into each chapter.
Step 4 – For each chapter, think about the remit of the chapter (i.e., what needs to be included) and the sequencing of this content, and develop suggested sub-headings to structure the content within each chapter.
Step 5 – Add notes on key quotes, readings, evidence, etc., for each (sub)section within the chapters, as well as any notes on gaps or other ideas – be this in terms of your literature, data, or analysis – to address later.

This planning should be a dynamic and iterative process that allows you to think creatively and update your plan as your thinking evolves. It can be helpful to do this using a series of large sheets of paper and a stack of (coloured) Post-It notes (or equivalent). Use the Post-Its to jot down your ideas on content, quotes, etc. (potentially using different colours for different 'themes' – theory, empirics, topics, etc.) and then slowly sift and organise these into different chapters (using a sheet of paper for each chapter) and then into sub-sections within each chapter. For instance, in thinking about a recent paper on gentrification and 'gener-ification' in Woodstock, Cape Town (Ah Goo et al., 2025), I found it useful to jot down ideas on themes, quotes, etc., on

Figure 3.2 Initial Organisation of Post-It Notes with First Thoughts on Quotes, Topics, Themes, and Key Literature Areas.

Post-It notes (Figure 3.2) and to then rearrange these to start to stitch the content together around key ideas and an initial structure (Figure 3.3).

Take your time with this process as the structure of your dissertation is vital to ensuring your ideas and argument are clear, logical, and convincing to the reader. Aptly for geographers, this planning matters at many scales, from the overall dissertation through each chapter and to individual paragraphs. Begin at the top-level, thinking through the focus of your dissertation before moving on to the sub-themes or topics that are the building blocks for this. You can then map these building blocks into different chapters and begin to plan the structure within each chapter. Moving through this process and towards a more sequential planning approach can help translate these ideas into a logical flow for your dissertation. Think through which ideas need to precede others, what information does the reader need in which order to understand the logic of your argument and the following content, and so on. Map this out to develop

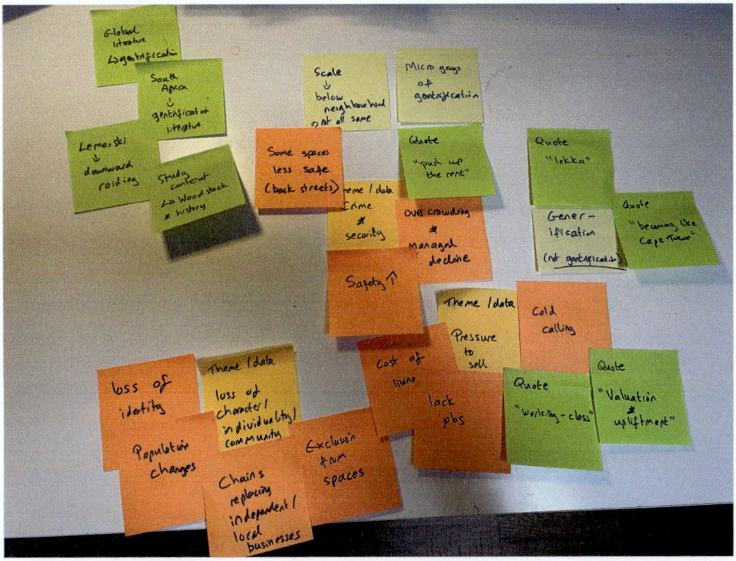

Figure 3.3 Re-organised Post-It notes as part of the article planning to identify how materials (quotes, literature, analysis) would fit together to develop a narrative and coherent argument.

chapter headings and sub-headings, ideas for different paragraphs and the order in which these need to come, and which ideas or evidence to bring into each section.

Many departments provide general guidance on the expected structure, typically a variation on the IMRAD approach (Introduction, Methods, Results, Analysis, and Discussion) with an anticipated structure of the following:

- Front Matter
- Introduction
- Literature review/conceptual framework
- Research Design and Methods
- Results and analysis/discussion (this content may be presented in one chapter or split into two – 'Results and Analysis' and 'Discussion' – see Chapter 7, *Step by step: Results and discussion* for a longer reflection)

- Conclusions
- Reference List
- Appendices

If you feel your dissertation would work better with a different, potentially more creative, structure – for instance if you are writing a cultural geographies dissertation or have used creative or innovative methods – then talk with your supervisor to check on the suitability of this. The following chapters in this book assume you will be using an IMRAD-esque structure and provides step-by-step guidance on each of these typical chapters. Once you have a plan in place, you are faced with the challenge of starting to write.

FURTHER USEFUL SOURCE OF ADVICE

The Royal Literary Fund was set up to support writers of all kinds. Their website contains a range of resources and guidance materials for students, including specific resources to support with the dissertation writing process which provide further tailored advice on specific skills and approaches in writing.

STARTING WRITING

Getting started on writing your dissertation can be daunting. For all the advice offered already on when and where to write, what happens if you are faced with the curse of 'writers block'? The cursor flashing in the top corner of that blank page on your laptop screen or the empty sheet of paper in front of you can feel almost impossibly debilitating. So, how do you get those first words out?

Firstly, consign any expectation that your writing at the outset needs to be good, let alone perfect, to the bin. Your first writing just needs to be something, anything: it can be intended for a specific part of your dissertation or may not have any clear target in mind in terms of location within your final document. The first hurdle to overcome is simply getting words out, with the shift to more directed and focussed writing coming slightly later. Useful approaches for getting started with writing can include:

Free-writing or stream-of-consciousness writing. Write anything (ideally directly or loosely connected to your dissertation but not necessarily) without worrying about spelling, punctuation, or grammar. Simply write for a short period of time, typically 3–5 minutes to begin with and then extending to 10 minutes if this is productive. As you begin to gain the confidence of getting words out, gradually shift towards a more focussed form of free-writing which is intended for a specific section of your dissertation, but still without paying attention to grammar and structure. Once your free-writing time has finished you should have a reasonable amount of text to work with. At this stage, you can start to edit and reshape, moving or removing content that you do not want to keep or is not useful, but developing and expanding on material that is and then correcting spelling, punctuation, and grammar.

Journal or diary writing. Keep a daily (or at least regular) journal related to your dissertation containing notes on ideas arising, questions emerging, or readings relating to your work. This may support you in the habit of frequent writing while building a library of materials to use in the substantive writing phase without the pressure of 'getting it right' or worrying unduly about the relevance of the material to your final argument.

Talk, don't write. Use your phone or a digital recorder to record you talking about your dissertation out loud – either to yourself or to a friend or family member. This might be in relation to the overall argument and flow of your work, or a more focussed monologue or conversation on a particular section or chapter. The more ephemeral nature of spoken words can help you express your ideas, and the style of communication means you can't dwell on getting the phrasing absolutely correct. You can then transcribe and edit these recordings later, working them into your overall dissertation as you go.

Map, don't write. Use a visual and creative approach to create mind-maps of sections of your dissertation, using Post-It notes and different coloured pens to identify and piece together ideas, concepts, topics, key sources/authors, quotes/examples/data, connections. Once these notes are on paper, you can begin to piece these into a structure and narrative, iteratively adding bits of text as you go, and eventually turning this into structured prose.

THE ART OF NOT BEING A PERFECTIONIST

In the early phases of writing do not worry about writing 'well' or 'perfectly'. A preoccupation with getting it 'right' runs the risk of you spending (wasting) hours honing a few sentences that then don't make the final cut of your dissertation. Get an early draft written without worrying hugely about grammar, spelling, and punctuation – nor even the word count (within reason). Once you have a substantive draft you can then focus on editing the work and honing your style and content.

As you do this, clearly differentiate any direct quotes from your own commentary or notes (to avoid accidental plagiarism) and use comment boxes to add notes that will help you understand why you have included content or if you think material is tangential or core, and questions you need to follow-up, etc. Treating your first draft as a living document – one that you are continually editing, adding to, removing from, making comments on, etc. – can be helpful in avoiding a preoccupation with perfection. That, and the acceptance that there is no 'perfect dissertation' only a 'good enough' one. An unfinished or unsubmitted dissertation can never be 'good enough'.

At these early stages of writing, it is imperative to avoid any aim for perfection – it is far easier to edit materials that are already drafted than to try and write perfectly the first time around. You will, therefore, end up with multiple iterations of you dissertation chapters, requiring you to develop good file management practices from the outset.

SAVING YOUR WRITING – AND GOOD FILE MANAGEMENT

No matter when, where, or for how long you are writing – be sure to regularly save your work. One option to consider is setting up an auto-save function on your computer to minimise the risk of losing significant amounts of work. This can be effective but can also mean you end up overwriting earlier versions and risk losing material in these drafts that you cut out but later want to add back-in and so on.

The best advice is to ensure that you save you work regularly using a strong file and version management approach to retain an archive of earlier versions of your writing in case these are useful later.

Common approaches to file and version management use either a version or date-based naming system (e.g., Intro v1, Intro v1.1, Intro v2, etc., or Intro 16 Aug 23, Intro 18 Aug 23, etc.). Not only does this allow you to identify which version of a file you are working on but enables you to return to earlier iterations of your work to find since-deleted content you want to re-insert or to track back to find a missing citation that was accidently removed during the redrafting process. You may also find it useful to have a separate document in which you save any sections of text that are cut from an early draft. These sections may never get looked at again, but occasionally you may realise you need to add a sentence, paragraph, or section in again having previously deleted it. Keeping an archive file of these materials can save significant time and stress compared to having to recreate content from scratch.

Alongside using a clear naming convention, make sure your work is backed-up: a lost, stolen, or broken laptop with the only copy of your draft dissertation is not what you want a week before your hand-in deadline! Storing back-up files on a thumb drive/USB stick is one option but is generally discouraged as it comes with added data protection and data management considerations and the risk of losing the device. The recommended approach instead is to save and back-up your files to your university filestore or secure online storage site (e.g., Googledrive, Onedrive). If you prefer working online on Google Docs (or equivalent), this has the advantage of being (auto)saved to the cloud along with an automatically updated version history. However, good practice here would be to save an offline version as a back-up in case of problems with accessing the cloud.

FINISHING WRITING

While much focus and emphasis is placed on getting started with writing, it is also important to know when to stop writing. For your dissertation there is a clear cut-off point – your hand-in deadline. Ideally, though, you should aim to have finished a first full draft of your dissertation a few weeks before this deadline to allow time to get feedback (from supervisors if allowed, but if not then from family and friends) and to revise and edit your work. In saying this, it can be hard to 'let go' of your dissertation – you will have spent months working on it and there will always be something else you *could* do. Remember there is no perfect dissertation – you just need to ensure it is good enough to submit with a sense that you have done justice to the time and energies spent on the research journey overall.

FINISHING THIS CHAPTER

This chapter has explored how the writing process involves multiple (often unacknowledged) decisions and challenges. There is no single correct way to start writing your dissertation, rather the chapter has provided you with a series of factors to consider in identifying the approach to writing that will work best for you – from thinking about when and where to write, to the use of milestones and rewards to both motivate and keep you on track through the writing journey. These strategies can help you reach the point where you feel able to finish writing and submit your work. The next chapter develops this further, providing advice to support you in writing well so that when you submit your dissertation you can be satisfied you have clearly communicated your key findings and argument.

4

GOOD WRITING

An overarching plan for writing your dissertation is important, but this needs to be coupled with good writing practice to ensure you clearly communicate your ideas and understanding. A clear structure, consistent 'red thread', correct spelling, syntax, grammar and punctuation, and an active voice all help convey your argument more effectively. This, in turn, means you can include greater depth and detail, provide more nuance, and demonstrate your contribution to knowledge with greater clarity and efficiency. In simple terms, if you can't clearly and explain your approach, theoretical framework, findings, and discussion you will undermine your project and struggle to demonstrate how meet the marking criteria (see Chapter 14 – *Remember your audience: Keeping your marker onside*).

You may be thinking that trying to change your writing style will be a tedious and pointless activity that takes much-needed time away from writing your dissertation. While improving your writing voice and style takes time initially, it is a worthwhile investment and should reduce the time spent later on editing drafts to meet word limits. Good writing and editing skills will also benefit your work for other assignments and with employability (see Chapter 15 – *What next? Employability and your dissertation*). Developing these skills is an iterative process and takes practice: the rest of this chapter provides advice on some of the easier ways to enhance your writing along with exercises to put these ideas into practice.

AN ITERATIVE PROCESS

Effective writing is an iterative process resulting in the concise and clear communication of a message. This requires multiple

DOI: 10.4324/9781003584315-4

rounds of (re)writing and redrafting – sometimes you won't notice you are doing this as you rework a sentence while writing. At other times this will be a concerted and focussed stage in the drafting and editing process. Regardless, it is a process of continual refinement and reflection on both *what* to write and *how* to write in terms of the flow and focus of content, the clarity, and precision of your argument.

Unsurprisingly effective writing takes time. To borrow a famous phrase from French physicist and mathematician Blaise Pascal, 'I have made this [letter] longer, because I have not had the time to make it shorter' (Blaise Pascal (1623–1662), '*Lettres provinciales*', *letter 16, 1657*). The point to draw from this quote is that it takes time to think not only about *what you want to say* but also *how best to say it.* A first stage in this process is to pause before you start writing and think about the aim and purpose of your writing (see Chapter 3 – *Writing your dissertation*). What are the key points to make? In what order should they go in? How do they build towards a larger argument and narrative? You don't need a firm and final answer to these prompts for a first draft but writing with purpose is useful and helps with the (re)drafting process. Remember, it is better to start writing earlier so you have time to repeatedly re-draft the content as necessary, and ensure your writing is not 'longer' as in Blaise Pascal's letter.

Good writing is a craft that takes time. Coherent, well-articulated sentences, paragraphs, chapters, and dissertations do not appear on the first iteration: you will need to hone your material over several drafts. This means you should

1 try not to worry about your first writing being perfect (or even good), just get words on paper and then edit later;
2 factor the time needed for multiple rounds of (re)drafting into your planning;
3 remember it is clear to the reader (your marker) when the writing process has been rushed.

To help with this process, you should think about writing with POWER. POWER stands for Plan, Organise, Write, Edit, Revise – and can be used both for the overall structure of your dissertation and for each chapter and (sub)section.

WRITING WITH POWER

Plan – Identify your topic and gather the information you need to write about this. Brainstorm your ideas and identify the (sub) themes to your topic. Use a mind-map or other approach to plan your writing (from the scale of dissertation to chapter, etc.). In preparing materials for writing, ensure that you have clear notes to use, including reference details and clear differentiation in your thoughts and ideas, summary notes, and direct quotes from sources.

Organise – Review and organise your ideas (from the Plan stage) and add in further levels of detail (in note form) on content for each section, ensuring the flow and narrative are clear.

Write – Develop your notes into full sentences and paragraphs, layering in further details and expanding on the ideas you have identified in your planning and organising. Don't worry at this stage if the syntax and grammar aren't accurate, the focus is on getting the overall narrative and key ideas in place.

Edit – Go back over your writing and correct for grammar, syntax, spelling, and other errors. While doing this ensure your ideas are clear and the narrative is logical – if not, then rework.

Revise – Revise your paper based on the editing process and re-edit as necessary.

READ TO WRITE

To ensure that you are effectively writing with POWER, it is useful to first pause and read – read the marking criteria, example dissertations, academic journal articles, books, newspapers, and so on. Some of these will have obvious direct relevance to your dissertation, others less so. By reading more widely, you will (perhaps unknowingly) be developing useful knowledge and ideas for your own writing in different ways:

1 Reading relevant academic and grey literature on your topic will develop your knowledge and understanding of the ideas, concepts, debates, and context that inform your dissertation.

You need to demonstrate your command of relevant literature – and reading is essential to this.

2 Reading examples of previous dissertations will improve your understanding of the requirements of your dissertation and how to effectively communicate and demonstrate that you have met these.
3 Across all forms of reading, you will expose yourself to differing styles of writing, helping you to think about examples of good and bad writing – how are ideas communicated? How do writers use narrative and other devices to effectively mobilise their argument? Granted, you won't write your dissertation in the style of a romance or murder mystery novel – but reading in general helps you understand how ideas can be effectively conveyed, and will help develop your vocabulary and writing confidence.

As outlined in more detail in Chapter 5 – *Step by step: Literature review* – you will use different reading strategies to engage with these different sources and resources depending on the purpose for which you are reading them. This wider reading will also help you to develop your own style and voice.

AVOIDING PLAGIARISM

By this stage in your degree, you will have heard repeated warnings about plagiarism and the penalties that will follow if you are found to have attempted to use another person's work and pass it off as your own. Plagiarism can be either intentional (you deliberately set out to cheat) or inadvertent and can occur if you

- use someone else's ideas or words without correctly citing the source;
- falsely claim an idea is new when you know it isn't; or
- submit someone else's work as your own.

Given the length of time you will spend working on your dissertation, there is an elevated risk of inadvertent plagiarism occurring. Some key tips to avoid this are as follows:

1. Use good note-taking habits and record all necessary referencing information about the publication in your notes on the source as you are reading it.
2. Clearly differentiate (using a different ink colour, clear quotation marks, etc.) between direct quotes (including page number for the source material), summary notes, and your thoughts and comments.
3. Ensure care and consistency in adding, deleting, and retaining citation and referencing details between different drafts of your work.
4. Add in your citations and references as you write (see Chapter 12 – *Step by step: Referencing and reference lists* – for detailed discussion of how to do this, and the benefits of using reference management software and 'cite while you write' functionality).
5. If possible submit your draft dissertation through TurnItIn (or equivalent plagiarism detection tool) to check for any accidental plagiarism prior to final submission. If any issues are identified you then have the opportunity to address these prior to final submission.

FINDING YOUR VOICE

This does not equate to attempting to sound 'academic' by using overly complicated language and jargon. We've all read 'those' papers – the dry, tedious, and confusing articles you struggle to make sense of. You don't want your marker to have that experience with your dissertation (see Chapter 14 – *Remember your audience: Keeping your marker onside)*. At the same time, you need to avoid an overly informal and personal writing approach as that can be read as lacking depth, detail, and professionalism.

Aim then for the middle-ground, suitably professional and clear but allowing your personality to come through as you make your argument (it is *your* argument after all) in an accessible and engaging manner. This should allow you to demonstrate your control and understanding of the theory, concepts, and evidence being presented, and to guide the reader through your interpretation and understanding of these materials and how they contribute to *your* argument. By having a clear voice in narrating your argument you can more clearly and powerfully mobilise your analysis and discussion, identifying your contribution to knowledge while conveying your passion for the subject.

RESOURCES FOR EFFECTIVE WRITING

To support your writing you will need supplies and resources. These might include the 'treats' to keep you motivated and reward yourself for progress (with apologies to my dentist, good coffee and sweets are my preference) as well as both IT and physical writing resources – a computer/laptop, peripherals and consumables, a dedicated notepad, block of Post-It notes, and set of pens for jotting down notes, brainstorming ideas, practicing free writing, and drafting and editing text. Additionally, a good quality dictionary and thesaurus can be invaluable.

A good dictionary (such as the *Oxford English Dictionary*) helps with spelling and ensuring you have got the correct meaning and usage of a word. For instance, if you were to write that "300 people were evacuated from a flooded building", this would mean 300 people were given enemas – not that they were taken to safety from the building. Why, you ask? While a building can be evacuated, to evacuate a person is to give them an enema. This may seem like an overly pedantic point to make but meaning and precision matter and you want to avoid inadvertently misusing language and communicating the wrong thing.

You can further strengthen your writing, vary your vocabulary, and add style to your work through careful use of a thesaurus. A thesaurus provides you with groups of synonyms for words, helping you to quickly identify alternate – and sometimes more apt or accurate – words to use. Thesauruses come in different styles – and degrees of comprehensiveness – with *Roget's Thesaurus* among the most well regarded. This may sound arcane but can make a tremendous difference in enhancing the readability of your writing, reducing repetition by varying terminology and wording, and providing more precise and specific language choices. If you don't want to invest in physical copies of these texts, your library will likely have online access to these or equivalent resources. At the very least, make use of your word processor's functionality for spelling, grammar, and synonyms. It may not be perfect, but it is better than nothing.

ADDITIONAL USEFUL GUIDES ON GOOD WRITING

Provost, G. (2019). *100 Ways to Improve Your Writing: Proven Professional Techniques for Writing with Style and Power*. Penguin Random House, New York.

Provost, G. (2020). *Make Your Words Work: Proven techniques for effective writing for fiction and non-fiction*. Crossroad Press, Hertford, NC.

Strunk, W. (2006). *The Elements of Style*. Dover Publications, New York.

Sword, H. (2012). *Stylish Academic Writing*. Harvard University Press, Cambridge, MA.

Truss, L. (2003). *Eats, Shoots and Leaves: The zero-tolerance approach to punctuation*. Fourth Estate, London.

EFFECTIVE WRITING – KISS

While correct grammar and spelling are crucial, they are but one component of effective writing. The top tip for overall effective writing? To *Keep It Simple and Straightforward* (KISS).

Don't believe me? Think of the academic journal articles you've read. Which have you found easiest and more enjoyable to engage with? Which have you resented reading? Which have you gained most knowledge and understanding from? The chances are, the most difficult ones to engage with have been convoluted, complicated, and jargon-filled. Meanwhile, the one's you enjoyed and gained more knowledge from are most likely those written in an accessible manner that – even when dealing with complicated ideas and theories – have offered clear explanations presented in accessible language. The lesson for your dissertation? Do not overcomplicate your writing. To help you achieve this, we can start with Strunk's (2006) advice on *The Elements of Style*: be clear and concise, be precise with your language, very sentence length and vocabulary, and use examples to keep your reader engaged. Alongside – and overlapping with – these ideas, there are seven 'rules' for KISS that will help to ensure your writing is accessible and powerful.

THE SEVEN 'RULES' OF KISS ARE AS FOLLOWS:

- Aim to use short sentences where possible (ideally between 10 and 15 words)
- Omit unnecessary words
- Use accessible words and terminology
- Use an active voice in your writing
- Use correct punctuation
- Write in an open, assertive, and reader-friendly style
- Have one clear argument or narrative – a 'red thread' – that runs throughout your writing

In a similar spirit, Helen Sword (2012) suggests that effective and elegant academic writing should make use of interesting (sub)titles, anecdotes, and examples, utilise opening sections that engage the reader through a challenging question or a story, use concrete nouns and active verbs, informative (and fun) visual content, and refer to a wide range of literature. She also suggests the use of humour – but having recently written a book on humour and politics (Hammett et al., 2023) in which were asked to use humour, I can attest that this is a very tricky style to deploy effectively and can be far more effort than it is worth.

Underpinning this advice is a common emphasis on the importance of using plain English. This is not to say you should dumb-down your writing or oversimplify your content but rather focus on your writing being efficient, accessible, and suited to your audience while clearly articulating your core argument.

EFFECTIVE WRITING – THE RED THREAD

Integral to effectively writing your dissertation is to keep a clear and consistent focus on your main argument – a.k.a. the 'red thread'. Your dissertation needs to weave together different literature, sources, ideas, evidence, and discussion around a key line of thinking that links everything together. This is the focus that connects your title to each chapter of your dissertation and builds a narrative that runs from your introduction to your conclusions.

THE RED THREAD

The red thread is a term used in many folkloric traditions that connects people who are destined to meet – in other contexts it denotes a theme or connection that links things together. In business and branding this might be between the wants or needs of an audience and the service or product being offered. In academia we use 'red thread' to refer to the core argument that runs throughout a piece of work and which holds together ideas from different literatures and empirical sources. In other words, the red thread is how you guide the reader through your logic and argument. To borrow from Greek mythology and the legend of Theseus and the Minotaur, the red thread is your way of guiding your reader through a maze of ideas and concepts to the successful outcome. Not, in this case, returning to King Minos having slain the minotaur, but rather delivering to your reader a well-informed insight and contribution to knowledge.

It is crucial to a strong dissertation to know what your red thread is and to convey this to the reader. While planning and (re)drafting your dissertation you should be mindful of the centrality of your red thread and use this to structure your work. To help map the connections between your red thread and different ideas and materials, it is useful to reflect on key questions about the structure, flow, and content of your work (see also Sword, 2012: 157–158):

- What is the main point and core argument of my dissertation?
- What are my research questions and how do they relate to the core argument?
- Why am I including this literature/concept? (How) Does it relate to my core idea or argument and have I made the connection clear? If this is not clear, then how do I make it so? Or should this content be removed?
- Why have I included these data? Is it purely to demonstrate a breadth of activities undertaken/data collected, or do they serve a purpose in building my main argument? If the relevance is not clear, how do I make it apparent? Or should this be removed for being a distraction from the main argument?

- Have I made the connections clear between my literature and data analysis? Does my discussion effectively draw on and cite literature either to support my argument or as a point of contrast and critical discussion?
- Does each section logically and clearly connect both to the overall narrative and to the immediately preceding and subsequent content? In other words, does each paragraph provide a logical narrative with clear connections within a chapter? And do the main ideas from one chapter connect to those in other chapters?

You may find it useful to use a more creative or visual approach to linking these answers to your red thread – for instance, taking a large sheet of paper and drawing a 'red thread' down the page and then adding notes on ideas, arguments, literature, data, and so on to this thread. In essence this is a variation or continuation of the planning process outlined in Chapter 3 – *Writing your dissertation*. This can help you to identify core and peripheral content and understand the sequencing of ideas within your work. This process can also help with identifying starting points to use in drafting your abstract (see Chapter 11 – *Step by step: Front matter*), as well as to identify potential sub-headings and paragraphs in your main text. Figure 4.1 provides an illustrative example of 'red thread' planning used during the drafting process of a paper on the intra-national politics of knowledge production in South Africa (Hammett et al., 2025). In this instance, the planning is focussed on main sections and content areas: this could be developed further to locate key quotes and more detail within each section.

PARAGRAPH PLANNING

In planning your dissertation, it is important to think about how each paragraph supports or speaks to the red thread of your work. Each paragraph should have a clear purpose and point, and should clearly follow on from and lead into the preceding and subsequent content respectively. Making these connections is crucial in supporting your reader to understand the logic connecting different parts of your work. As a writer, it is easy to assume the reader knows our thought process when they do not – a key part of planning your work is to ensure you explain the connections and logic to your reader.

Figure 4.1 Example of 'Red Thread' Planning (Using the Example of an Article on the Intranational Politics of Knowledge Production in South Africa).

This means it is important to think about the paragraph as the unit of composition and how to structure each paragraph and its fit to the overall narrative. Each paragraph should focus on one topic, opening with a topic sentence which is developed in the body of the paragraph, before a final sentence that emphasises the important idea to take forward into the following material.

Some paragraphs more clearly and self-evidently flow from one another, but you should signal the connections. This might be through discussion of a series of viewpoints or approaches (First, second, third; One approach, a second approach, etc.), or through using the same word/idea/focus in the closing sentence of one paragraph and start of the next. Alternatively, it may be in offering contrasting opinions on a theme or idea between paragraphs, or in answering a rhetorical question posed in closing one paragraph and being addressed in the next. Regardless of which approach(es) you use, the key message is to ensure you signal to your reader how each paragraph links to the surrounding content and builds a clear narrative.

HOW LONG SHOULD A PARAGRAPH BE?

There is no formal limit on the length of a paragraph and it can be useful to use paragraphs of varying lengths within your writing. However, unless you are writing online content for websites, it is advisable to avoid single-sentence paragraphs as these become disjointed and isolated very easily. At the same time, if your paragraph is notably longer than half a page of text (single spaced) this is probably too long for a reader to easily follow and understand. In this situation, look to split the material into multiple shorter paragraphs, each with its own focus and purpose.

WRITING SENTENCES

Effective paragraphs rely on clear, focussed, and well-crafted sentences. Each sentence should serve a purpose and contribute to your overall narrative. There is no one-size-fits-all rule for what a 'good' sentence must look like. But there are common traits. Think about text you enjoy. What about the sentence length? Short sentences can get annoying. They feel really quite disjointed. Have you noticed this yet? Bored of just five-word sentences?

A good essay or dissertation will contain a variety of sentence lengths and types, making your writing easier to read. Changes in the length of a sentence inform how your reader engages with your ideas. You can convey more depth and meaning through a variety of short, medium, and long sentences, as well as by including rhetorical questions and other devices. (For more advanced ideas on this, see Gary Provost's (2020) *100 Ways to Improve Your Writing*.)

As a general guide, however, you should be aiming for a sentence length of 10–15 words, with one idea per sentence (and perhaps one related point). The use of shorter sentences will generally help your reader follow your thinking while avoiding multiple sub-clauses that can cause confusion. While more complex sentences can be used to convey more detail and information, it is important not to over-complicate things. Rather than trying to squeeze multiple (sub)statements into a sentence, it can be more effective to break content into multiple sentences. The following examples illustrate different sentence types and the related strengths and weaknesses of these. A sentence containing multiple statements and subordinates is known as a *compound-complex* and is difficult to write effectively and challenging to read. Other main types of sentences are *simple* (containing one statement), *compound* (two statements joined together), and *complex* (a dominant statement with subordinates).

SENTENCE TYPES

Simple sentence = Three politicians spoke out on international poverty yesterday.

Compound sentence = Three politicians spoke out on international poverty yesterday and called for an increase in development aid spending by 20%.

Complex sentence = Three politicians, one from the Conservative party and two from the Liberal Democrats, spoke out on international poverty yesterday and called for an increase in development aid spending by 20%.

Compound-complex sentence = Three politicians, one, an outspoken Cabinet Minister from the Conservative party and two, both backbenchers, from the Liberal Democrats, spoke out on international poverty at a press briefing at 11 am yesterday and called for an increase in development aid spending by 20%.

Rewritten across multiple sentences = Three politicians from multiple political parties spoke out against poverty at yesterday's 11 am press briefing. An outspoken Conservative cabinet minister and two Liberal Democrat backbenchers united to demand an increase in development aid spending by 20%.

As these examples demonstrate, simple sentences are very readable and effective at conveying core information in an accessible manner. Compound sentences allow you to convey further detail and depth while retaining a good level of readability. Complex sentences are, as the name suggests, more complicated and include greater detail through the use of subordinate statements. They remain relatively readable, but a series of successive complex sentences can rapidly become confusing. These sentences are best using sparingly and in conjunction with other sentence types. Compound-complex sentences, meanwhile, contain lots of information and detail but are difficult to read and can easily be misunderstood and cause confusion. It is generally advisable to avoid compound-complex sentences and present the relevant content through multiple shorter sentences to enhance readability.

Linked to this (and at the risk of repetition), do not fall into the trap of trying to impress your reader with overly complex writing and verbosity. More effective writing, especially for dissertations, should focus on 'telling a simple story elegantly'. In the early years of my academic career my mentor offered the following – invaluable – advice: always aim to 'write so a first-year undergraduate can understand your argument'. This advice is just as apt for a dissertation as an academic journal: if you over-complicate your writing, your marker will find it harder to follow your thinking and be more likely to misunderstand your argument. You will also increase the risk of syntax and grammatical errors creeping in. In other words, KISS – Keep It Simple and Straightforward. There are some common writing mechanisms you can use to achieve this, from removing redundant words to using an active voice.

AVOIDING UNNECESSARY WORDS

Long, complex, and unwieldy sentences can often be simplified by the removal of unnecessary and redundant words. The previous sentence would have conveyed the same message with fewer words: "Overly long sentences can be simplified by removing unnecessary words". If I can save seven words with a simple edit to one sentence, imagine how many words you could remove across your dissertation simply by using more concise wording. Culling unnecessary words not only helps with meeting the length limit for your dissertation, but it also supports a more active and authoritative voice in your writing by removing non-comital or hesitant terms (e.g., "not often" becomes "unusually" or "infrequently" depending on context) and phrases (the classic example in a dissertation being a variant of "This dissertation will aim to explore" which can be rewritten as "This dissertation explores").

Commonly used redundant words include: can, therefore, thus, often, not, frequent. This is not to say you should never use these, but to identify when they are necessary. Other oft-used redundant words arise from using phrases where a single word would suffice, or where a negative is being used, for instance:

The question as to whether	Whether
Used for travel purposes	Used for travel
She is a woman who	She
This is a policy which	This policy
But also	And
Is not clear	Is unclear

The key focus here is on ensuring your writing is clear and definite. In practice, you can see the difference made with some simple changes to sentences in the following paired examples.

Pair One:

The policy had a number of unexpected negative outcomes.
The policy was flawed.

Pair Two:

There is general agreement among commentators that the financial markets hold too much power over the economic policy decisions of sovereign governments.
Commentators agree that financial markets hold too much power over sovereign states.

Which sentence in each pair is more concise? Clearer? Most effective in communicating the key message? As these examples demonstrate, wordiness (using more words than necessary and the use of overly complicated language [Provost, 2020]) can be distracting and unhelpful. We can all be guilty of over-complicating our writing, often in an effort to sound serious, important, and knowledgeable. In reality, these efforts often backfire by making the argument difficult and confusing to follow. To counter this tendency you should focus on using an active voice in your writing.

ACTIVE WRITING

What is meant by an 'active voice' or 'active writing'? In simple terms, it is a more assertive, concise, and direct way of writing that communicates your message more clearly. Whereas passive writing is long-winded and has the subject on the receiving end of the verb, active writing has the subject performing the actions of the verb. This may all sound a little abstract, so consider following examples:

Passive = The rain caused the cactus to flower.
Active = The cactus flowered in the rain.

Passive = Spending on education was decreased as a result of the Structural Adjustment Policy.
Active = The Structural Adjustment Policy resulted in a decrease in education spending.

In both examples, an active voice provides a more concise, clearer, eloquent, and assertive sentence. This is achieved by ordering the three key parts of the sentence (the subject, the verb, and the object) to produce an active verb. To so this, we put the subject (the doer of the action) before the verb (the action), which is then followed by the object (the thing being subject to the action). For instance, if your dissertation was on nature-based flooding solutions you might have a sentence that said, "The increase in natural space along the riverbank reduced downstream flooding". Here, the sentence follows the order of subject-verb-object. If we reordered this to become a passive sentence ("Downstream flooding was reduced by the increase in natural space along the riverbank"), we use more words and make the sentence less reader friendly.

Conveying your argument clearly, succinctly, and powerfully will help illustrate your command of the literature you are using, the strength of your critical thinking and analysis, and the red thread of your argument. Using, where possible, the active voice is crucial to this – and will also help you avoid excess wordiness – as the following example illustrates.

Wordy: The provision of high-quality children's recreational facilities in major urban centres is identified in this research as a major priority for the improvement of accessibility and multi-generational utilisation of central business district areas.

Clear and concise: Providing play space for children in town centres would increase the number of families coming into these spaces.

The second version uses a more active voice and conveys the argument more clearly and effectively. It is not just in academic writing the overly complex writing causes confusion. The Plain English Campaign has compiled a wealth of examples of unnecessarily long-winded and complex writing, and suggestions for plain English equivalents. It is worth looking at their website for an illustration of how common ineffective writing is and how simple changes to active writing can make a radical (and positive) difference to readability. No matter how active your voice, if it is incorrectly punctuated you may find the whole meaning of your sentence is lost.

PUNCTUATION, GRAMMAR, AND SYNTAX

It would take an entire book to provide a detailed review of and guidance on punctuation, grammar, and syntax (see Truss, 2003). The focus here is instead confined to a few common mistakes and errors that creep into dissertations.

Change in tense – typically from past to future and back again. Usually this happens when someone has reused and incompletely edited a part of their dissertation proposal for their dissertation (e.g., methods). This is a clear giveaway to the reader that the writing process has been rushed and there is a risk of self-plagiarism that needs to be checked. By the time you are writing the final draft of your dissertation you are not 'going to do X', you 'did X'. With the exception of highlighting potential areas of future research (when you will need to use phrasing such as 'Building on the findings presented here, further research could examine'), your dissertation should be in the past tense.

Incomplete sentences – these occur either when words have been removed or missed out, or the subject or verb of the sentence is missing. These can be a sign of rushed writing or of incomplete editing.

Incorrect use of punctuation – this can change the meaning of a sentence, obscure the subject and object of the sentence and confuse the reader. As the fictional example below illustrates, the presence or absence of a comma or two can fundamentally change the meaning.

The Moldovans said the French government were at fault for Montserrat's descent into war.

The Moldovans, said the French government, were at fault for Montserrat's descent into war.

Punctuation errors can readily creep into a dissertation during the editing process. While unlikely to cause a diplomatic incident, they can result in significant misunderstandings and confusion as to the focus and meaning of your writing. To help ensure you are using punctuation appropriately and consistently consider the following advice.

Commas. There are six times at which to use a comma:

to separate clauses in a sentence;
to separate the introductory part of a sentence from main subject;
to separate the closing part of a sentence from main subject;
to separate either non-essential or essential information from the subject of the sentence (usually used in pairs);
to separate instructions or interjections from rest of sentence;
to separate adjectives or items in a list.

If you notice you are using multiple commas within a sentence to separate multiple types of information or interjections this can be a signal that your sentence is becoming too long and overly complicated (a complex-compound sentence – as mentioned earlier). It may be useful to consider whether all the information is necessary and whether you can more clearly communicate the content in multiple, shorter sentences.

THE 'OXFORD COMMA'

Debate remains as to whether or not to use a comma after the penultimate item in a list – the 'Oxford comma'. There is no definitive rule on this so check your departmental style guide and follow any guidance offered there. If no guidance is given, then decide on whether you will use the Oxford comma or not and be consistent throughout.

Apostrophes are a major source of confusion and poor writing. Follow a few simple rules and you can avoid glaring mistakes.

Use an apostrophe to

- show when letter(s) have been left out (informal, not commonly used in academic writing – examples include can't, shouldn't, won't);
- to indicate ownership/possession (e.g., Shiobhan's car or Mukhove's perspective).

Semi-colons should only be used to join two independent clauses in a single compound sentence. For instance:

> Staeheli's articles are engaging; they are written in using accessible language.

In essence, if the two clauses could be written separately as complete sentences (i.e., Staeheli's articles are engaging. They are written in accessible language.) then the correct punctuation in a compound sentence is a semi-colon. While you may catch many of these errors as you are writing and editing, ensure you have time for a thorough proofread before you submit your work.

YOU CAN FIND USEFUL FURTHER RESOURCES ON COMMON WRITING MISTAKES BY SEARCHING ONLINE FOR THE FOLLOWING:

The Writing Centre, University of Wisconsin-Madison – 12 Common Errors

Oxford International – Common English Grammar Mistakes and How to Fix Them

Grammarly – 30 Common Grammar Mistakes to Avoid

Hume Centre for Writing and Speaking, Stanford University – Top 20 Errors in Undergraduate Writing

The Robert Gillespie Academic Skills Centre, University of Toronto-Mississauga – Ten common Grammatical Errors and Conventions in Academic Writing

PROOFREADING

Proofreading should not be an afterthought. It needs to be an integral part of your writing process. Poorly presented work looks unprofessional, can lead to significant misunderstanding, and can directly affect your grade. As part of this process be sure to pay attention to consistency throughout your dissertation – in punctuation (e.g., use of Oxford comma or not), in capitalisation and formatting, in the use of UK or US spellings (e.g., color or colour, -ise or -ize), and so on.

The process of proofreading takes time and it is often tempting to rush – but try to resist this urge. You can see that I rushed the proofreading of one of my first articles if you look carefully at the opening few pages of 'Disrespecting teacher: the decline in social standing of teachers in Cape Town, South Africa' in the *International Journal of Educational Development* (Hammett, 2008). Fortunately, this proofing oversight (a repeated word) did not result in any misunderstanding, but it does not make a good impression.

Proofreading your own work can be tough – you may feel like you know the work so well you don't need to or can't bring yourself to read it in enough detail – so see if family or friends can help with this. One obvious option is to ask a course mate to proofread your dissertation in return for you proofreading their work.

KEY TIPS ON PROOFREADING ARE AS FOLLOWS:

1. Factor this into to your writing schedule and do not rush it.
2. Buddy-up with a course-mate and proofread each other's dissertations – you can also then help each other by highlighting places where the meaning is unclear or confusing.
3. Don't proofread when you are tired or distracted – you need to concentrate.
4. Don't try to proofread too much at a time – it's easy to start skimming sections and missing mistakes. If you find yourself doing this, take a break and come back to it refreshed.
5. Try proofreading a short section on-screen and on paper to work out which approach is more effective for you and use this approach for the rest of your dissertation.

PRACTICING EFFECTIVE WRITING

As reiterated throughout this chapter, effective writing is a skill that takes practice to hone and develop. This final section contains a series of short activities that put into practice some of the ideas discussed in this chapter.

A Rewrite the follow sentences in an active voice, removing redundant words where possible:

There are many ways in which each of us can make our writing style better. Indeed, these few sentences here could possibly be rewritten to produce a better end result in terms of the readability of the text.

____

____

____

____

B Rewrite the following sentences into an active voice, ensure the use of simple language and correct any punctuation errors:

1 The argument made by Hammett (2008) is flawed.

____

____

2 The car engine was changed by Josephine on Tuesday morning.

____

____

3 The television set was switched off by David. It's ability to receive programmes was poor in bad weather.

____

____

4 The ethical conundrum faced by the researchers emanated from their efforts to circumvent the local topography of cultural expectations surrounding the contentious issue of remuneration.

____

____

5 Beinarts proposition; that the moon is made of cheese, was subject to derision by the audience.

____

____

6 The Oscar winner said Ken Loache was inspired by a toy mouse.

____

____

C Rewrite the following 200-word paragraph in 100 words:
A review of recent academic literature has indicated how the presence of tall skyscraper buildings in the centre of Indian cities is a sign of social and economic progress. Researchers, however, caution that increasing preoccupation with the aesthetics of the built environment risks planners and policy makers overlooking the potentially devastating impact of these buildings upon the environment and upon marginalised, poor communities. The development of these buildings, often linked by elevated walkways and private roads, almost universally displaces and excludes street-traders, the homeless, and other impoverished community members. At the same time, the resources required to build and maintain these buildings can potentially result in the reduction of local water resources and increased levels of pollution. Urban planners, researchers at the University of Venda argue, face an imperative to refocus their vision and energies on to the synergistic, symbiotic development of multi-sited, multi-audience built urban landscapes that provide accessibility to multiple strata of society and account for the impaired physical

mobility of the many mobility-impaired and destitute citizens forced by economic circumstance to adopt street-based survival strategies. In mobilising this argument, the researchers point to the success witnessed in recent urban planning and regeneration policies evident in southern Angola.

CHAPTER SUMMARY

Building on the previous chapter's focus on approaches to writing your dissertation, this chapter has offered more focussed guidance on ways of writing well. Beginning with the idea of writing with POWER, the chapter has then introduced specific strategies for accessible and active writing and identified common errors in writing and how to avoid these. A series of optional exercises have then given you the opportunity to practice and refine your writing skills. The next set of chapters move on to guiding you through the process of writing each chapter of your dissertation. The guidance in this and the previous chapters will be of relevance and use throughout. As noted in the introduction, the following step-by-step chapters are sequenced to map onto the more common order in which dissertations are written (not read) – starting with the literature review.

5

STEP BY STEP

Literature review

Your engagement with existing, relevant literature in developing your dissertation proposal and questions means you should already have the foundations for your literature review. While you are likely still reading materials related to your dissertation, the literature review chapter is typically (one of) the first you will begin to write. The purpose of your literature review is to provide the detailed synthesis and critical review of existing literature that informs and frames your dissertation. While you cannot include everything published of relevance to your research question, you need to demonstrate your understanding and command of the key debates relating to your project. This *step-by-step* chapter outlines the purpose of your literature review chapter(s), offers advice on how to structure your literature review, reflects on how to review literature in ways that will support this writing, and addresses some common questions.

WHAT SHOULD A LITERATURE REVIEW DO?

Your literature review is the framework of ideas within which you develop and situate your research and argument. It should demonstrate your understanding of the state of knowledge within the field and show how this informs and interacts with your research. To do this effectively you need to convey your command of these materials and articulate an argument from these. You need to *review not report* the literature; your review needs to have a purpose and argument, and not simply summarise what other people have said.

DOI: 10.4324/9781003584315-5

Think of film or book reviews. They do not simply report the narrative and plotline but give you the critics' views on the strengths and weaknesses of the film. Similarly, your literature review needs to provide your critical take on the literature you have read and the ideas, concepts, and theories therein.

Reviewing not reporting means you are critically synthesising – drawing together – multiple sources to develop an argument and provide a rationale for your dissertation. It is not a space to simply demonstrate all the readings you have done by describing (reporting) each source. You need to mobilise ideas from, and your analysis of, the sources to build *your* story and *your* argument. As such, you should ensure each paragraph and sub-section contributes to this overall narrative and is relevant to the material that follows.

Building on the brief indication of the main argument of your dissertation in your introduction (see Chapter 10 – *Step by step: Introduction*), your literature review provides the substantive evidence of your understanding of the terrain that you are contributing to. This is where you:

1 Demonstrate an understanding of the field of work you are engaging with. This should establish your detailed knowledge of the realm of your work, and the strengths, weaknesses, gaps, and opportunities therein. This will provide the foundations to your research and analysis. As the adage goes, your work is 'standing on the shoulders of giants': the literature review is where you set out those 'shoulders' and how you will build on them. The narrative you develop here will underpin your entire dissertation with the literature also then used to inform your discussion and analysis (see Chapter 7 – *Step by step: Results and discussion*).
2 Signpost awareness of other important debates in the literature that are relevant but not directly applicable to your research. These materials can be strategically signposted in the opening section of your literature review, demonstrating that you have engaged thoughtfully and widely with the topic, and made an informed and justified decision as to your specific focus. One classic formulation to this involves a sentence or two summarising the following:

> Debates on topic X have explored considerations of factor P (relevant citation(s)), factor Q (relevant citation(s)), and factor R (relevant citation(s)). Alongside these discussions, factor T has emerged as one of the most important aspects to understanding how topic X.

The second sentence then leads to a deeper discussion of the literature and materials relating to your dissertation focus. This approach simultaneously shows your wider knowledge while then funnelling attention to your question and aims.

3 Set out the conceptual material you will use to inform your analysis. You should draw on ideas, materials, and sources from your literature review in your discussion. This means you need to ensure the key ideas and concepts you use in your discussion feature in your literature review. And vice-versa – the key ideas in your literature review should be integral to your discussion and analysis. If they aren't, then there is a mismatch that you need to address. For instance, if your dissertation is on barriers to social mobility in Humberside and your literature review focusses on economic factors but your discussion and analysis emphasise social capital, then you should revisit your literature review and engage further with debates on social capital while paring-back the economics-focussed literature.

4 Develop an argument within the literature review. Your review should have a purpose or objective – by the end of this section you should have conveyed to your reader not only the key ideas or background but also what *your* thinking is on a topic.

5 Highlight any (relevant) gaps in literature and understanding, and use this to justify and frame your research as making a contribution to knowledge.

GIVING YOUR LITERATURE REVIEW A STRONG TITLE

Using generic chapter titles is a missed opportunity for you to signpost the focus and flow of your argument. Your literature review is no exception to this. Think about how authors of academic journal articles or books use chapter and section headings to situate their work and guide you through the connections between ideas. Using thematic or conceptual (sub-)headings for sections that are reviewing and synthesising literature can help you to indicate the focus and/or argument you are making and to guide your reader along your 'red thread' much more easily than the overly abstract and generic chapter title such as 'Literature Review'.

STRUCTURING YOUR LITERATURE REVIEW

As with each chapter, you need to begin with an introductory paragraph or two to set up the rationale and remit of the chapter. For your literature review this would include signalling the broader debates in which your work is framed, signposting the specific focus of the literature review, and outlining the chapter structure. Ideally this should connect to your introductory chapter, signalled through phrasing such as "To address the questions outlined in the introduction" or through reference to the ideas contained in the introduction. By signposting clear links between different chapters you help your reader follow your core narrative and logic – this mechanism is really helpful with maintaining the red thread of your work. Alongside this signposting, you should also outline the purpose of and rationale for the literature review, previewing the argument you are making and conclusions you reach within the chapter.

The substantive content of your literature review should then be structured to provide a clear evaluation of relevant debates. There are various approaches you can utilise for this structure, including thematic, theoretical, chronological, hypothesis/challenge orientated, and methodological. A *thematically or theoretically organised* review will set out and work through relevant debates grouped into thematic areas or theoretical approaches (for instance: differing schools of thought on systems for rewilding projects, or different conceptual approaches to theories of citizenship). A *chronology-based* review will track changing debates and ideas over time, critically reflecting on how these have evolved before highlighting current gaps in thinking. This can work most effectively where there has been a clear development of approaches and thinking – for instance in the emergence and evolution of approaches to women and gender in development, or the influence of a particularly seminal paper that reshaped a field, such as the work on capabilities by Amartya Sen and Martha Nussbaum. Be careful with this approach to avoid simply recounting a timeline of debates and ideas – remember you need to provide your critical analysis of these debates and build your argument from these.

A *methodology-based* review would critically examine literature using different methodologies in thinking about or researching a topic and the key findings within and across these different approaches. This might be in the emergence and utilisation of more creative methods in cultural geography for instance, in the different methodological 'turns' within human geography, or the emergence of new technology and techniques for remote sensing, geographical information systems, or laboratory-based analysis. In presenting material here, you would need to develop a critical discussion of how these different techniques have contributed to understanding and knowledge, and build towards a rationale for the research design and methods you are using in your dissertation. Meanwhile, a *challenge/hypothesis orientated* review would explore literature that underpins different perspectives, hypothesis tested, or challenges addressed on the topic of interest and develop a narrative based on these factors.

Whichever approach you adopt, do not try to cover too many concepts or ideas – remember to keep a clear focus on your 'red-thread' (Chapter 4 – *Good writing*) and mobilise a core argument. To support this ambition, do not think of your literature review as a single continuous block of text. There will be various facets to the debates in the literature and opportunities to bring different ideas and perspectives into conversation with each other. To clearly communicate this, break your literature review into focussed sections using sub-headings to signpost the structure and flow of the chapter. If you are not sure how to start this process, look at the journal articles you are using and consider how they use sub-headings to inform you about the content that follows and steer your attention to specific ideas or concepts. A further advantage of using sub-headings within the chapter is that these provide natural points at which to provide brief introductions to and conclusions or summaries of key points or ideas you are working with, and how these connect both to your overarching argument and to other sections within the review.

Finally, your literature review should conclude with a brief overall summary of your analysis of the existing literature, reiterating key ideas and gaps, and signposting how your work engages with these. You might also introduce your research questions here, particularly if they have been derived from the literature review (see Chapter 2 – *Supervisors, support, and common questions* – for more on where to locate your research questions, and Chapter 10 – *Step by step: Introduction* – for tips on presenting research questions)

FROM REVIEWING THE LITERATURE TO WRITING THE LITERATURE REVIEW

You should have started reviewing the literature well before you start writing your dissertation. It is, though, a process that will continue during the writing-up process. Keeping up to date with and integrating newly published literature into your dissertation can be a key strength to your research. As you should be well advanced with the process of searching for and reviewing literature – and because this book is focussed on the writing of your dissertation – this section focusses on translating your reading and notes into your literature review.

HANDLING A LAST-MINUTE PUBLICATION ON YOUR SPECIFIC TOPIC

If a book or article on your dissertation topic, question, or case study is published just before your hand-in deadline no not panic. Your reader knows that you will have been working on your project for months – they will not expect you to redesign or rewrite your dissertation due to a new publication. If you have time, then you can strategically use this in your introduction to illustrate why your topic matters, and (if possible) incorporate into your literature review.

It can be helpful to set yourself a deadline after which you will not search for new literature. You do not need to detail this anywhere in your dissertation, but it can minimise anxiety about newly published works late in your dissertation journey.

Firstly, be focussed in the coverage of your literature review. Hopefully you have been strategic in your use of search terms and identification and selection of sources to read. If you are still reviewing literature then be selective in your search strategy –identify appropriate combinations of key words to narrow your focus to more relevant documents, use date-limits to focus on recent papers and debates, or restrict selections to those with relevant geographical focus.

For dissertations based on systematic literature reviews you should have kept a log of which search engines you used, the search terms employed (and relevant wildcards or limits such as dates or publication outlets, and modifiers such as 'AND', 'OR', and 'BUT NOT') and in what combinations, the numbers of sources returned for these searches, and the process through which you narrowed your search and made final selections (including how many articles were removed as duplicates or deemed to not be relevant). You would then be expected to include this detail in your methods chapter. While you are unlikely to be required or expected to provide this information for a standard review chapter, it is nonetheless useful to understand the process you used and identify any gaps or omissions you might have to address.

PRESENTING A LITERATURE REVIEW IN A SYSTEMATIC LITERATURE REVIEW-BASED DISSERTATION

If your project is a systematic literature review, then separating this initial literature review section from your systematic review analysis and discussion can feel awkward as your data is also the literature. In this situation you should include a short literature review providing a succinct signposting and summary of the overall state of the field that identifies major debates and themes to provide an overall contextual framing. You then use this to justify the focus of your specific systematic literature review.

No matter how selective and strategic you have been in your search and review process, you still need to exercise discretion in writing your review chapter. Think about why each source is relevant and worth including in your review: what is the key detail, information, or insight you want to draw from this source? Does it provide a new theory or insight? Does it offer important contextual information or background data? Does it bring a new perspective into debates or identify the need for work to engage with a different approach, perspective, or method? Is it further evidence of the extent of engagement on a topic, critique, or debate? Understanding the relative importance of a source to your review will help you determine whether you include a longer discussion of the piece, use as a supporting citation to bolster a point made using other sources, or leave out altogether.

Key sources that provide the core ideas or framing for your work will need detailed outline and discussion, supported by critical engagements with other sources that may support, adapt, critique, expand, or develop the original ideas. As with the skill involved in reading for breadth versus reading for depth, your literature review needs to balance these two imperatives. You need to make strategic use of a breadth of sources to show you understand the research horizon and landscape (breadth) but quickly focus onto a particular set of themes, debates, and issues and demonstrate your detailed, substantive engagement with these (depth).

As with other essays you will have written, common priorities to consider in writing you literature review include:

- Writing with purpose and having a clear focus and argument. Signpost this from the outset and maintain throughout the chapter, making use of relevant sub-headings as well as signposting and concluding sentences to paragraphs and sub-sections to guide your reader.
- Integrating sources. Don't individually summarise one source, and then another, and then another in discrete paragraphs. Place them into conversation and synthesise key ideas to build and support the argument you want to make based upon different sources' arguments, data, assumptions, oversights, and relations to other work in the field.
- Your literature review should consist of reputable peer-reviewed academic sources – not social media posts, news media stories, and so on. Depending on your focus, you may also draw in 'grey' literature such as government reports, civil society organisation

or think tank reports, and similar – but be aware of the political agendas and positionality of these sources and think critically in how you use them.

- Not all academic sources are equally good or reputable. It pays to be aware of the quality of the journal or publisher that you are drawing work from. There are growing challenges from predatory journals and publishers (offering impossibly fast 'peer review' times and publication timelines in return for 'author processing charges') and the phenomenon of 'highjacked journals' (where a company takes over a lapsed domain name or creates a new site using the logo and branding of the legitimate journal as a cover to publish poor quality works). If you are unsure of the reputability of a (purportedly) academic source you can check with your supervisor or university librarian if necessary or check Scopus or Bealls' list for relevant information. You can also use various academic search platforms (from Google Scholar to ORCID) to check the profile of authors.
- Avoid 'shopping lists' of citations. While it can be useful on occasion to include multiple citations to make a single point or illustrate the breadth of work on an idea, your literature review is not a demonstration of how many sources you have read. You need to apply the ideas and knowledge gained from the sources and show your critical thinking based on an *appropriate breadth* of sources.
- If you don't draw on a source (directly or indirectly) in your review, then don't include it as a citation. Similarly, if you do draw on a source – to inform your thinking, as a direct quote, via paraphrasing of the content – then cite it.
- Ensure that the literature you are using is up to date. While you often need to include some older sources to demonstrate your understanding of seminal works and foundational thinking, you also you need to show you are engaged with recent debates. A literature review that is based solely on articles from the 1990s or early 2000s suggests either nothing of relevance has been published in 20–30 years (very unlikely) or that you have failed to engage with recent debates in the field (more likely). This will be picked up as a key weakness to your dissertation.

COMMON INDICATORS OF WEAKNESSES IN A LITERATURE REVIEW ARE AS FOLLOWS:

While the previous section focussed on how to write a good literature review, this section briefly summarises some of the more common shortcomings and weaknesses in literature chapters that you should seek to avoid.

Paragraph(s) that rely upon one source at a time. Sometimes this is necessary to set out the ideas, theories, or concepts put forward in a seminal paper. More often it is a sign that you are just reporting rather than reviewing content, suggesting that the literature is leading your work rather than being used to support and inform your thinking.

Very few cited works. On rare occasions this may illustrate that there genuinely is very limited pre-existing work on the topic of your dissertation. More commonly it is a sign that the review has been rushed and you have a limited awareness and command of existing debates.

The literature review is repetitive. This can arise from a mixture of descriptive summarising of existing works and a lack of clear focus and structure to the literature review. This can indicate the review is rushed and incomplete, or that you are trying to mask a limited amount of effort and engagement.

NARRATING YOUR ARGUMENT

As you develop your literature review you need to assert your voice through your analysis and argument. This is key to highlighting your confidence in moving beyond reporting and to critically synthesising and evaluating existing literature. This narrative should not be written in the first person (e.g., "I think that") but expressed through phrases such as "Advocates for approach B claim X. However, criticisms remain" or "Limits to A have resulted in a recent turn to C". In making these claims, be careful not to overclaim or overgeneralise and be cautious of definitive wording such as "No research has been done on" or "The absence of work on". Using more circumspect wording such as "While much research has focussed on X, there remains relatively limited engagement with Y" still conveys that there is a niche for you to address but avoids the risk of your marker saying 'you claim there is no work on X, but here is a long list of references on just this issue/topic'.

In developing your argument you need to be cognisant of and articulate the strengths and weaknesses of different works and approaches. This does not mean that you need to give equal weight to differing views nor to all viewpoints. Do not fall into the fallacy of 'balance' being equal space for all viewpoints – you can (and should) dedicate your focus to the argument you want to make in your dissertation. In so doing, you need to draw out key ideas and debates, evidencing your recognition of and rationale for adopting one approach over another as part of this process.

PARAPHRASING AND DIRECT QUOTES

A good literature review will likely contain some direct quotes from the source literature. But be careful in how many you use: too many direct quotes and the reader will feel you are not in control of the literature but being led by other people's thinking rather than your own insight. Be selective, use quotes sparingly and their effectiveness will increase as you have more space to provide your analysis and discussion of these materials.

It is also wise to be cautious of using extended quotes in the literature review. Not only can these quickly eat into your word limit, but they do little to demonstrate your critical understanding of the materials you are working with. Where you do include longer quotes, be sure to provide your analysis or commentary so the reader can understand why you have felt important to include.

Shorter quotes can be useful when used to highlight specific ideas, definitions, points, or findings within the literature. To be effective, these quotes need to be integrated into a narrative that makes clear why they are there and what you want the reader to take from them. Overuse of quotes – of any length – in a literature review can quickly lead to a sense that the literature is leading the way or is in control of your argument, rather than you having a clear argument that is support by the relevant sources. Prioritise using your own wording to communicate your understanding of the key ideas from the literature – being sure to include relevant citations.

CHAPTER SUMMARY

Your literature review is a substantial piece of work, one that underpins the conceptual clarity and overall strength of argument of your dissertation. This chapter needs to clearly articulate your command and understanding of existing debates and build a strong narrative from this that flows throughout the rest of your dissertation. The literature mobilised here will intersect with and inform both your research questions and your research design and the methods used in your dissertation, as the next chapter discusses.

6

STEP BY STEP

Research design and methods

The research design and methods chapter is commonly viewed as one of the easiest to write as much of the content is based on an account of what you did to collect and analyse your data. While much of this is relatively straightforward to write, a good methods chapter goes beyond simple descriptive recounting of what you did. It needs to explain, justify, and critically reflect on of the decisions made and approaches used in data collection, preparation, and analysis. The rest of this *step-by-step* chapter guides you through the purpose and typical content of a research design and methods chapter (simply 'methods chapter' hereafter).

WHAT IS THE PURPOSE OF THE RESEARCH DESIGN AND METHODS CHAPTER?

The underlying purpose of your methods chapter is to outline *how* you did your research *and why* you did it this way. A good methods chapter will

1 demonstrate a clear, logical, and appropriate research design for your project;
2 justify the approach taken and decisions made in designing your research and collecting your data;
3 explain and justify how you have prepared and analysed your data;
4 acknowledge limitations and challenges to your research and how you considered and addressed these;

DOI: 10.4324/9781003584315-6

5 reflect on (as appropriate) ethical, risk, and positionality considerations in your research design and practice;
6 demonstrate that your dissertation is based on robust, reliable, trustworthy data, and analysis.

In summary, your methods chapter evidences the foundations upon which your research is built. If the underpinning of your research design and data collection is or appears flawed and ill-considered this will undermine the reader's perception of the quality and reliability of your data and analysis, and of your dissertation overall. Therefore, while your methods chapter may be seen as an 'easy' section to write it can easily become both a missed opportunity and the Achillies heel of your dissertation if it is rushed.

STRUCTURING YOUR METHODS CHAPTER

As always, the chapter introduction is key to connecting your methods chapter to other material in your dissertation and should signpost the structure and focus of the chapter itself. Connections to previous chapters can be made through links back to research questions (if already included) – common phrasing for this includes "To address the research questions outlined in xx" – or to the big debates and ideas set out in the introduction and/or literature review – for example, "Amidst growing concern with the impacts of climate change on river hydrology in the Conwy Valley this dissertations employs". You might also use this opening paragraph to briefly introduce and justify the overarching approach taken to your dissertation as framed by a particular epistemology and therefore how this frames your methodology and choice of methods.

The actual structure of the substantive sections of your methods chapter will vary depending upon sub-disciplinary norms and decisions as to where to locate your research questions (see Chapter 2 – *Supervisors, support, and common questions*) and/or contextual background. If you have not already included your research aim(s) and questions in either the introduction or literature chapters, then these will likely be the first substantive section of the methods chapter. This information needs to be provided to your reader before you explain your methods as the reader needs to know what it is you are trying to find out about before being told how you have tried to do this.

The next section in your methods chapter will then provide the (case study) contextual background to your study (if not already included elsewhere – which may be as a short standalone chapter either before or after your literature review depending on the natural flow and logic, or integrated into your introduction if only limited detail is required). Wherever this content is located, your reader needs to have sufficient knowledge and understanding of the context for your research – the 'where' of the focus – to be able to see how these conditions inform your later analysis. The 'where' will also likely inform the 'how' of your research. Providing this contextual detail here is useful in framing your choice of methods and reflections on the data collection, collation, and analysis process.

The next sections of your chapter will then outline the methods used and rationale for these choices, specific details of what was done, along with reflections on key limitations and how these were overcome or accommodated, and – perhaps more in qualitive-methods-based dissertations – considerations of positionality and ethical concerns. As outlined throughout this chapter, it is important not to fall into the trap of treating your methods chapter as a routinised obligation to be met by simply mechanistically listing what you did. Good methods chapters provide insight into why you have designed and conducted your work in this particular way and reflect on this process and your learning from this.

USING SUPPORTING LITERATURE

The purpose of the methods chapter is to demonstrate your understanding of and ability to design and successfully deliver a research project. To do this, you need to demonstrate that you understand the thinking behind research design and practice. This means you should draw upon (and cite) relevant literature to support your justification for your research approach and design, choice of methods, and to inform your reflective discussion of research practice. Weak methods chapters tend to have few citations of existing literature, with any materials cited in formulaic and generic ways. Stronger methods chapters provide specific details on the decisions made and actions taken, using supporting literature to inform the justifications offered and to reflect critically upon strengths, weaknesses, and challenges of the research activities undertaken.

JUSTIFYING YOUR OVERALL APPROACH

Offering a brief overall justification of your research approach is a useful section to include early in your methods chapter. The aim here is not to describe in huge detail your understanding of the -ologies (ontology, epistemology, methodology) but to outline and justify the broad rationale for your research approach. If you are using a particular methodological or analytical framework such as Critical Discourse Analysis or Water Associated Disease Index or another tool to frame and inform your research design then you should outline and briefly this here .

THE -OLOGIES

While your methods chapter does not need a textbook-style recounting of the different -ologies involved in research design, it is nonetheless helpful to know what these are and how they matter for your dissertation. Some dissertations will benefit from an explicit (but brief) justification and connection being made between your ontological position and methodological choice. More often, this can be provided in a more implicit manner by your justification of methodology and methods in conversation with your overall research design.

Ontology – theory of being. In other words, what we assume exists.

Epistemology – theory of knowledge. In simple terms, what we can know about what exists and how we know what we know. Some common epistemological standpoints include the belief that there is no objective truth and meanings are always produced through interactions (*constructivism*), that meaning is created but without being rooted to a specific object (*subjectivism*), or that meaning exists separate from consciousness (*objectivism*).

Methodology – theory of method. Put another way – our understanding and justification of the suitability of the methods used in empirical research (remember – this is not the same as your method(s)!)

This should be a fairly short section to set up the big-picture for your research design and methods decisions. Care is needed, though, to ensure it is effective. If rushed and approached in a formulaic manner, this can read as a weak descriptive summary of note from a research design textbook. A strong outline section will instead succinctly demonstrate that you understand how and why the approach you have taken is justified and logically connects between data, methods, questions, and the understanding being sought.

Following from this, you should proceed with an overview of your data collection and analysis (methods used, sample size, analysis approach, etc.) and justification for these decisions. This is key in demonstrating that your work addresses a worthwhile topic or opportunity, has a reliable and appropriate research design, and has developed robust data and findings through appropriate and rigorous data collection and analysis techniques. Doing this requires you to outline and justify the methods used, describe clearly and effectively what you did, and demonstrate how this allowed you to address your research questions.

CONTEXTUAL BACKGROUND

You need to provide sufficient contextual detail for your reader to understand why the location of study is important and relevant to the questions being asked, and how this informs the data collection and analysis. The decision on where to position this detail depends on the style and type of project being delivered, and how much detail is required. Where limited contextual information is needed this might be absorbed into the introduction chapter or as a short sub-section within the methods chapter. Where more substantive details are needed, this may work better as an extended section within the methods chapter or as a shorter standalone chapter (either between the introduction and literature, or literature and methods chapters).

Consider these hypothetical situations. A student is completing a dissertation on gender and the use of public space in Leicester, with the primary focus on gender and use of public space. Leicester simply provides a convenient location for the study. In this instance, a relatively limited amount of background information on Leicester is needed and this could either be absorbed into the introduction or methods chapters. Contrast this with a student completing dissertation on dark tourism and post-genocide reconciliation in Rwanda. Alongside conceptual materials outlining theories linked to destination imaging and dark tourism, the reader would require a relatively well developed contextual framing to the case study – as this is integral to the questions being asked. A short standalone chapter or (more likely) a lengthier section within the methods chapter would be better suited to providing the necessary contextual background detail.

In thinking through what to include in this section the key question to consider is 'what contextual information does a reader need to be able to understand this research and the data and analysis presented?' For instance, a dissertation on British electoral geographies does not need to include average temperature or rainfall, just as for a dissertation on riverine erosion rates the reader does not need to know the religious composition of local communities. Identify what descriptive detail is needed – quantitative and/or qualitative – to demonstrate the relevance of the context to the research along with sufficient background understanding that the reader can logically follow your analysis.

This information does not need to be textual only – contextual details and background can be effectively communicated using maps and other visual materials (see Chapter 8 – *Step by step: Presenting data*). As ever, do not simply include images and information for the sake of it – ensure it contributes something beneficial to your work and referenced in the main text. For instance, a dissertation on contrasting perceptions of public space in Cape Town, South Africa, and Manchester, UK, would benefit from annotated maps of the two cities alongside descriptive details on socio-economic and other factors in the relevant neighbourhoods to underpin the analysis that follows. Likewise, a dissertation on soil contamination in the Blackmoor Vale in Dorset would benefit from visual materials that locate this area for the reader, highlight sampling sites, and any key contemporary or historical natural, built, and industrial landscape features.

STRUCTURING THE CHAPTER

The exact flow and content of your methods chapter will vary depending on the approach and methods used but would typically include description and justification of the following:

1. Case studies or units of focus. This content would introduce the (geographical/locational) focus of your work and explain why this is relevant/appropriate to your questions (but see earlier section in this chapter on contextual information). This section may benefit from the inclusion of maps, photos, or other visual materials to help 'set the scene' for your reader (see previous section).
2. Specific methods used. Here you would be expected to show that you understand why this/these method(s) are appropriate for your research. You need to illustrate that you understand how the method 'works' and the type of data or information being collected or elicited, provide details on the number of interviews conducted, samples taken, etc., and the strategy used in this process. In doing this, you need to communicate why this/these method(s) in particular are useful – what do they allow you to gain insight into and how does this relate to your research aims/questions?
    a. For qualitative methods such as interviews you should explain how you recruited participants (including any sampling strategy), how many (interviews) were done, how long they lasted, where they were done, and a general outline of the focus of the questions. You do not need to list of your interview questions but it may be helpful to signpost in broader terms what topics the questions covered. (N.B. Some departments require interview schedules to be included as an appendix to the dissertation, so be sure to check your regulations).
    b. For scientific or laboratory-related methods, you might expect to include details on your sampling frame and strategy, where and when samples taken, and using what equipment, how samples were stored and processed, and so on. You might find it important to acknowledge the origins of any scientific methods used and what their purpose is

(i.e., don't assume your reader is an expert in the exact field of your dissertation), the steps taken in the process of measuring, storing, processing, and analysing samples, how variables were recorded and/or assigned, etc.

c If your dissertation is based on secondary data – such as climate data or census data, land use data, vulnerability datasets, Afrobarometer, or other largescale survey data – then you should outline the data source(s) used, how you accessed these, a brief justification of the validity and relevance of the source(s), how you 'cleaned' the datasets, and checked for internal consistency (e.g., comparability of questions across different Afrobarometer surveys), and so on.

3 Analysis approach used. Whether for quantitative, qualitative, or scientific data, you will have analysed your data. You need to demonstrate how you did this (whether in relation to coding frames, regression and other statistical analysis, laboratory techniques, and so on), any software used in this process, and the rationale for this approach. This would also encompass details on any programming written or equations and similar used as part of your analysis (in relation to programming, you might summarise this here and include the code itself in an appendix; see Chapter 13 – *Step by step: Appendices*), the process used to construct GIS maps and other outputs and how you validated these. For qualitative data-based dissertations, your approach to coding – including what platform you used (e.g., NVivo) and how you developed your coding frame – needs to be detailed in this chapter.

LISTS OF INTERVIEWEES

If your department demands that you include a list of interviewees – and assuming you have conducted interviews – then you must do so. If this list is optional, then think about its usefulness. If you would be including it just to show how many people you spoke to then this is a waste of words as you should be providing details on the number of interviews in your text anyway. If you are using the list to provide brief contextual information on different respondents that is useful to the reader, then this can be a helpful but not essential addition (in such cases, the table might be better suited as an appendix). A more effective approach can be instead to provide a very brief amount of contextual information about interviewees when presenting quotes from or discussions about these individuals (see Chapter 8 – *Step by step: Presenting data* for more guidance on this).

If you are including detail on participants in your dissertation – whether in a table or within the substantive text – ensure that you adhere to the requirements set out in your ethics approval regarding anonymity. Unless there is specific agreement to use someone's actual name or employer's name, etc., then use pseudonyms, generic job or company descriptions, etc. Once you have drafted the content, be sure to double check that it won't be possible for someone to piece together details within your dissertation to identify who a respondent is. Each extra piece of information provided about a person increases that risk, so be diligent to ensure you do not accidently reveal too much.

A good methods section will draw on existing literature to justify approaches used and decisions made, highlighting any weakness and how these were addressed. Ultimately your reader needs to know what you did (and thus be able to repeat the work using the same methods – the idea of replicability) and to understand why you did the research in the way you did and to see that the data collection (and analysis) approach was appropriate, justified, reliable, and robust. You do not need to rely solely upon extended prose to convey all this information – tables or diagrams can be useful mechanisms for communicating some of this methodological detail. This might be through a flow diagram of the steps involved in collection and analysis of data or in the production of graphical outputs; a table outlining definitions of categories used in your analysis (e.g., land cover classes/types and soil group types) or summarising data type and methods of calculation; or a matrix showing the connections between research question, method, and data (e.g., Table 6.1).

Table 6.1 Example Matrix Illustrating Connections between Research Questions, Methods, and Data

Research Question	*Method/Data Collection*	*Data Source*
Research Question 1	Interview – expert	Experts from relevant organisations
Research Question 2	Interviews – community Media analysis	Community activities Community members Lexis Nexis
Research Question 3	Interviews – community	Community activities Community members
	Participant observation	Community meetings Protest events

ETHICS AND POSITIONALITY

Your dissertation should have undergone an ethics approval process. In some departments, you may be required to include your ethics application or approval certificate as an appendix. Nonetheless it is worth noting in your methods chapter that your research was conducted in line with the approval granted by your ethics review process (assuming this was required). For dissertations using qualitative and/or personal data your methods chapter should include a brief reflective and critical commentary on ethical challenges or dilemmas faced and your response to these (your 'ethical response-ability'; Kraftl et al., 2021; Hammett et al., 2022) and/or a brief outline of how you delivered on ethical practice in your research.

Similarly, there is typically a *de facto* expectation that qualitative-methods-based dissertations will include consideration of the researcher's positionality. Within qualitative research this matters as there is an understanding that meaning is made from the researcher's analysis of the data rather then the data themselves: and this analysis is subjective. In other words, your personal and social beliefs and life-experience with shape the questions asked and engagements with people during the data collection process as well as how you interpret the data. Your positionality statement will thus set out key aspects of your identities, experiences, and privileges, and consideration of how these may have shaped the research and findings.

With both the ethics and positionality sections of the chapter it is important not to approach these as routinised, box-ticking obligations. If you do so, the result will be a generic and simplistic commentary on a common ethical consideration or a recounting of a few protected characteristics without substantive reflection. Such an approach signifies to your reader – your marker – a lack of care and attention and suggests a lack of appreciation of how ethics and positionality underpin both the data collection process and quality of data collected. These brief sections are often a missed opportunity to easily show your reader that you have a critical awareness of the research process. The first step to providing a more useful commentary on these issues in relation to your dissertation is to think about why a particular ethical consideration or positionality factor mattered *for your dissertation*. You should then reflect on what this meant for any changes to or limitations to your data collection, with your discussion of this concern then informed by and supported with relevant literature.

LIMITATIONS

No research is without its challenges or limitations – whether arising from a smaller than anticipated sample size, a restriction on available laboratory resources limiting the number or types of analyses run, the granularity of a secondary dataset, or simply the time available to complete your dissertation. A crucial part of your methods chapter is to acknowledge the challenges and limitations encountered and how you mitigated these. It is generally inadvisable to list every single challenge, dilemma, or obstacle faced. Instead, identify one or two key limitations or challenges and provide a more substantive reflection on how these impacted your data collection, analysis, and/or findings, and how you responded to these (if possible). The purpose of this section is for you to demonstrate that you are a reflective researcher, that you can critically appraise the relative successes and shortcomings of your research practice and learn from these experiences. You do not want to cast doubt on the reliability of your approach, data, and analysis, so be selective but honest in the detail you share here.

DISCUSSING CHANGES TO METHODS

Research plans rarely survive the data collection phase without some kind of alteration or adaption – whether a change in the number of interviews conducted, the need to move a sampling site from one location to another, or the change in the exact dates used to analyse historic climate patterns. As a result, it is likely that the exact details of the methods and research design outlined in your dissertation proposal will differ to an extent from the reality of your fieldwork. If these changes are covered by existing or updated ethics approval this should not be a problem and you do not necessarily need to declare these differences in your dissertation.

Whether you opt to include details of changes to methods is an individual decision. For most minor changes, it is simpler to *focus on what you did* rather than discussing what you did not do. For instance, if you completed 12 rather than 15 interviews or gained 380 rather than 400 survey responses, you do not need to discuss this relatively minor discrepancy in detail. Rather, focus on what you *did* and why you did this.

If you made more significant changes to your research design due to unexpected factors, or made minor changes that had significant impact on the outcome of your dissertation these can be worth explaining and justifying in more detail. The key is not to cast doubt on your original plans, but instead to articulate how and why your approach changed – either as a predetermined contingency plan for a key potential failure point or due to unexpected challenges – and to show these changes ensured robust data collection. Regardless of whether changes to your methods and research design were made, this section needs to clearly explain to the reader what you did and why. Do not fall into the trap of assuming the marker has intuitive knowledge of what you did – you need to tell this part of the story.

CHAPTER SUMMARY

The research methods chapter of a dissertation is an oft-missed opportunity. Too often students approach this chapter as a formulaic and descriptive piece of writing, resulting in their underselling their own work and missing out on key aspects of the marking criteria. It is relatively simple to move beyond description and into critical reflection in the methods chapter, using existing literature and personal experience to justify decisions and adaptations made, and to offer substantive reflections on ethics, positionality, and other limitations and considerations. Spending a little time developing these extra insights ensures your reader moves on to your results and discussion with confidence in the underlying data while demonstrating how you meet relevant marking criteria (see Chapter 14 – *Remember your audience: Keeping your marker onside*).

7

STEP BY STEP

Results and discussion

Your results and discussion chapter(s) are your opportunity to demonstrate both the quality of your data and your ability to synthesise and analyse these materials to make an informed, insightful argument. This argument should draw on and inform existing literature, making it clear that your discussion critically connects your data to previous work and narrating what your contribution to knowledge is. This *step-by-step* chapter outlines how you might approach structuring the content and flow – from introduction, through substantive content, to closing summary – and mobilise an effective argument. Before exploring how to do this in detail, you need to know whether you are writing two chapters ('Results' and 'Discussion') or one combined chapter ('Results and discussion').

RESULTS AND DISCUSSION: TOGETHER OR SEPARATE?

Whether your results and discussion should be presented in separate chapters or not will firstly be informed by your module guidelines and marking criteria. If these explicitly specify one approach or the other, you should adhere to this guidance. If there is no such instruction or the phrasing is more open (e.g., 'where appropriate, results and discussion are well separated') then the choice is yours, albeit informed by sub-disciplinary and/or methodology-related norms.

Qualitative methods-based dissertations tend to combine results and discussion to avoid unnecessary repetition. This style is rooted in an understanding that data and analysis are subjective, with the

DOI: 10.4324/9781003584315-7

meaning derived from the data through the subjectivity of the analytical process. In other words, it would be misleading to present the data as objective 'facts' or truths when they have been shaped by the data collection and subsequent analytical processes. Instead, you should anticipate interweaving your data, analysis, and discussion along with use of supporting literature to develop your argument.

Laboratory and quantitative data-based dissertations typically separate results from discussion. This provides for the presentation of data (results) as objective measurements in a clear, descriptive fashion – often including the use of maps, graphs, tables, and so on – accompanied by initial analysis. This analysis is then advanced through broader discussion in a separate chapter which mobilises your interpretation of the data and analysis in conversation with existing literature. With this approach it is crucial to ensure that your results chapter communicates your key data and the analysis of these materials only. Synthesis and discussion of the meaning and importance of these results then follows in the discussion chapter, which refers back to and builds on the results and analysis chapter. The discussion chapter then provides a broader critical review of the findings and interprets their meaning in relation to the research questions and existing literature. Here you want to be making it clear that you have answered your research questions while also adding to existing knowledge.

RESULTS AND DISCUSSION – TITLE(S) WITH MEANING

As discussed for your literature review chapter, using a thematic, concept, or topic-based chapter title for your results and discussions chapter(s) can effectively signal to the reader what the materials that follow are focussed on. This can also be useful for you in writing up as you can consistently sense-check that your content fits to the chapter title *and* the key argument of your dissertation rather than straying into tangents and sidelines. As well as using descriptive chapter titles, clear and informative sub-headings should be used to scaffold the flow of the chapter and the specific findings being raised at different points. Sub-headings here should be short, focussed, and primarily descriptive of the following sub-section.

For instance, in a dissertation on the ethics of care the results and discussion chapter might be titled "Geographies of care in practice" with sub-titles then on "Caring about", "Caring for", "Being cared for", and "Caring with". This set of titles reminds the reader that the results are coalesced around a central theme (care) with the sub-titles then providing a rationale for and structure to consider and explore the data and analysis on specific aspects of care. In quantitative or laboratory-based dissertations, where results and discussion are separated and often based on hypotheses, (sub)titles that summarise the key focus can also be used but may be less creative. For instance, for a dissertation on air pollution and sporting events the results chapter might be contain sub-titles focussed on different air-quality tests or on different testing locations, as well as information on wind speed and other relevant factors. The discussion chapter might then be structured on a per question or per hypothesis basis – while it can be quickest to simply list as Hypothesis 1, 2, 3, etc. (or even as H1, H2, etc.) in the sub-titles (and therefore contents page) this does not help the reader to a great extent. Instead, include some basic descriptive information to remind your reader of your key questions or hypotheses.

INTRODUCING YOUR RESULTS (AND DISCUSSION)

Your results and discussion need a short introduction that briefly signposts the focus and flow of the chapter, and the importance of the argument being mobilised. This should provide a very succinct teaser of your main finding(s) and allude to how you advance existing knowledge. It is a key opportunity to reiterate the red thread of your work and how your results and discussion contribute to this. Even if your results and discussion are not overtly structured by your research questions you can use this opening paragraph to highlight how you are addressing your research questions through the different thematic (or other) sections of your chapter.

STRUCTURING YOUR RESULTS AND DISCUSSION

Following from a short introduction, you need to think carefully about the flow of the rest of the content and ensure there is logic to this. Some of the more common overarching approaches to structuring this material are:

1 *By research question or hypothesis*. This approach allows you to clearly communicate your response to each research question. This can clearly signal to your reader how you have met key aspects of the marking criteria but may result in disjointed or repetitive content depending on the intersections or overlap of your research questions, analysis, and whether you are using the same data(set) to answer more than one question. This approach can also limit the potential to draw out cross-cutting themes or to mobilise broader arguments, feel formulaic, and encourage descriptive answers. If you adopt this structure, be clear in differentiating the data used for and argument made relating to each question. In some dissertations there may be a very clear alignment between data source/type and hypothesis so structuring your chapter by question or hypothesis makes sense. In other dissertations this alignment may be far less clear and a different structure – by cross-cutting theme for instance – may be more appropriate.
2 *By (cross-cutting) theme*. It may feel somewhat counterintuitive to structure your discussion using themes or ideas that cut across your research questions, but this can be effective in communicating your wider understanding and argument. This approach can work well in cases where cross-cutting themes have emerged during analysis and speak across multiple research questions. In these cases, be sure to signpost how you are addressing your research questions (potentially at the end of the chapter or in the conclusions).
3 *By data type/source/case study*. This approach is sometimes utilised when using multiple methods or working with several data sources, and the focus is on understanding the key findings emerging from each 'source'. This can work effectively if these then tie closely to a specific research question(s) or to support a two-stage process of per-source analysis that then feeds into comparative analysis between data type/source/case study. While this can be effective, care is needed to ensure the case-by-case content is not overly descriptive (in the discussion/analysis) and the level of repetition between sections is minimised or strategically mobilised to discuss and analyse similarities or differences.

Making this choice can be one of the more challenging and time-consuming parts of the process. Give yourself time to plan once you have a clear sense of your data, analysis, and key arguments, and think about the flow, focus, clarity of argument, and logic of connections being drawn. It might be useful to develop a few possible outline plans using mind-maps or equivalent approaches before finalising your choice. Even if you do revert to your first idea, this process should help you think through the logic of why this approach works, how to handle your data, and how to narrate your argument.

Crucially, with whichever structure you use, the key priorities are to:

1 Clearly communicate your findings, analysis, and discussion in a logical and accessible manner.
2 Present a robust and logical argument that is fully supported by the evidence included.

REMEMBER THE LITERATURE

Whether your results and discussion are integrated or separate, you need to support your discussion and analysis with existing literature and debates. If your dissertation has separate results and discussion chapters, the results chapter is unlikely to have many (or even any) literature sources being cited. This section would be focussed on reporting *your* data and analysis. The literature would then be more effectively integrated into your discussion chapter.

Across both approaches, a common weakness in dissertations is the failure to draw connections between the literature review and the discussion. When this happens, the discussion can read as a series of unfounded assertions. The key here is to weave ideas and debates from the literature through your discussion and to use this to inform your discussion. While using the literature be sure to prioritise *your* argument and analysis – avoid the temptation of framing your findings to fit to the existing literature. Linked to this, avoid starting paragraphs claims from the literature or quotes from other authors (this gives the impression that you are following their lead) and instead lead with your data and discussion, using the literature later in the paragraph to support your argument.

To help effectively integrate literature into your review, it can be useful to think about instances where your results

1 corroborate existing debates in the literature;
2 provide further evidence of the validity or applicability of existing debates and advancing these into a new area;
3 offer an alternative to or contradict existing literature and findings.

Where you can make these connections, draw the literature into conversation with your data and explain how you understand the relationship between the two. Do not leave it to the reader to have to make this connection – you need to demonstrate *your* critical thinking skills by making these links. In other words, your analysis matters. Crucially, this analysis needs to be rigorous and informed – failing to draw literature into your analysis and discussion can result in your claims being weak and partial, or being viewed as assertions rather than informed analysis.

YOUR ANALYSIS MATTERS

As discussed in Chapter 8 – *Step by step: Presenting data* – your reader needs to see your data. This means you should not simply summarise the findings but present the empirical evidence *and* your analysis to the reader. This is imperative to demonstrate that you can – and have – critically analysed the data and made meaning from it. This is vital in terms of delivering on key marking criteria, but also in ensuring that your reader follows your line of thinking and argument – if you don't present your analysis of the data, your reader may interpret it in a different way which can cause confusion. In practical terms, this means that each table, graph, quote, image, etc., needs to be accompanied by your commentary and explanation.

To help structure this, consider the following:

1 Why is this piece of evidence included?
2 What is this piece of evidence telling me about the bigger story?
3 How does this piece of evidence relate to existing discussions and literature?

How you structure and narrate your analysis and discussion will vary according to methods and data. For interview-based dissertations your reader needs to be able to both read the 'voices' of your participants, meaning you need to include direct quotes from your interviews (or similar methods), and see your analysis of these data. This does not mean you include a quote and then paraphrase it but is your explanation of the meaning you are making from the data. This is vital in connecting the quote(s) to your main argument while demonstrating your ability to analyse data effectively. For quantitative and scientific methods, you will still be drawing out the meaning and importance of the data and your analysis of this, albeit in a more objective manner.

WRITING YOUR ANALYSIS

As outlined in Chapter 4 – *Good writing* – effectively communicating your ideas requires a clear flow and concise writing style. Effective analysis will make direct use of empirical materials to illustrate and substantiate your argument, while critical use of existing literature will support your argument and interpretation. This does not mean you simply state as fact an individual quote or data point and then add a sentence supporting this claim from the literature. Rather, you need to reflect on the materials presented and explain what these mean and your evaluation of the evidence. A common weakness to results and discussion chapters is a tendency towards description over analysis – of simply stating what happened or was said, and assuming the reader will make meaning from this. In writing your analysis it is *your* responsibility to do analytical work for your reader.

To help you be consistent in applying a critical lens in writing this content you need to prioritise explaining and justifying over simply providing. Rather than stating what is known you should explain why this matters. Instead of saying what was done, explain why this was done. Rather than listing connections, evaluate the relevance or extent of these connections. Instead of stating a theoretical approach, demonstrate the relevance of this and how it helps explain your analysis. To help with this process, you might use a form of Socratic questioning – a process that helps you to assess materials (from the literature or your data collection) and reach your own conclusions about these. Addressing the following series of six prompts should help you to build a critical understanding and argument:

1 Clarification – How does the material relate to your topic? What else do you need to find out about your topic?
2 Assumptions – What is the claim in the source material based on? Can this be checked or verified?
3 Evidence – What is the proof/evidence provided to support the claim? What examples are used? Are these reliable and appropriate?
4 Viewpoint – What other approaches or viewpoints might there be? What are the limitations and strengths of the viewpoint provided?
5 Implications – What are the likely outcomes or consequences?
6 Question – Why does this question matter? What other questions could/should have been asked?

Your responses to these prompts may not provide clear-cut and definitive answers but can be used to help articulate the (dis)connections are between sources (and data) and how you understand the meaning and importance of these.

WRITING YOUR ANALYSIS – ACTIVITY

Take one short piece of your data and write this up (to a maximum of one page) using three different styles that you have seen in the literature you have read for your dissertation. To do this, take three articles of relevance to your work and think about how the author(s) of these pieces have used different structures, narrative devices, phrasing, etc., to communicate their results and analysis.

Ideally, try to use a range of styles – for instance, one adopting a more creative writing style, another using a narrative approach, and another using a thematic approach.

Testing out different styles of writing your analysis should allow you to experiment with these approaches, find your voice and style, and help you to think through different ways of structuring and presenting your findings.

CHAPTER SUMMARY

Strong analysis and discussion explain not only the data themselves but also why they matter. Your results and discussion chapter(s) are commonly among the more substantive sections of your dissertation, and carry tremendous importance in marking criteria. Spending time to develop and present this content in a nuanced, sophisticated, and critical manner is vital. In doing this, you need to connect your findings – and argument – to existing literature. Be sure, though, to foreground your voice and analysis, and to use literature to support but not lead the discussion. As you do this, it is essential that you explain the logics and connections to the reader – do not rely on or make your audience piece together the threads of your thinking. You need to guide them, using your red thread to logically take your reader on a journey through your dissertation. Integral to this journey is your data, and you need to ensure this is presented in a clear and compelling manner to effectively underpin your analysis.

8

STEP BY STEP

Presenting data

Data is at the heart of your dissertation, whether modelling ice-melt on Alpine glaciers, gauging community perceptions of coastal erosion and habitat management policy in Norfolk, or analysing changing geographies of voting patterns in Scotland. Your data provide the evidence for your analysis and conclusions. They underpin your entire dissertation. It is crucial that these materials are well presented and clearly explained, demonstrating that your argument is based on evidence rather than supposition and conjecture. This *step-by-step* chapter discusses key considerations in effective data presentation and provides guidance on common approaches to and methods for presenting both quantitative and qualitative data within geography dissertations.

EFFECTIVE DATA PRESENTATION

Effective communication is key to your entire dissertation – including precision in presenting your data. Effective presentation of data includes clarity in differentiating between data, literature, and your analysis or discussion. Attention to how, where, and how much data you present to your reader is key. This latter point is perhaps the hardest for many students to reconcile with. Having spent weeks or months collecting and analysing data, you naturally want to demonstrate the fruits of this labour. This can result in the common mistake of simply shoehorning as much data as possible into a dissertation. This becomes a distraction and limits the clarity and depth of your main argument. You need to tell a clear story that

DOI: 10.4324/9781003584315-8

is based on robust – and robustly analysed – data. To do this, you need to include *sufficient* data to support your argument and show that your research and analysis are reliable and meaningful. This does not mean you should include every single quote, graph, table, or other piece of data you have.

On the flip side, failing to include enough data is equally problematic: without sufficient evidence to support your analysis, the reader cannot be sure of the reliability of your arguments and findings. You have to show you have an evidence base for your discussion and conclusions. The key questions to ask yourself are as follows:

1 Have I included sufficient evidence to adequately support each point I am making?

   If not, then you should look to add additional relevant evidence and data.

2 Have I included data that are irrelevant to my key points?

   If so then you either need to make a relevant point from the data that supports your core argument or, if you cannot do this, then either delete it or consider moving to an appendix as additional supporting material.

3 Have I included data that is unnecessarily duplicating findings already evidenced?

   Sometimes you need to triangulate findings so having data to show this is useful – perhaps this is from interviews and newspaper coverage. But if you are using multiple quotes or data tables that provide the same information unnecessarily consider removing the duplicate content to save words and ensure clarity in your discussion.

In short, data are key to your dissertation but you should only include those which develop your argument. Focus on including data that are relevant to and required to substantiate your argument, communicating this in a way that is readily accessible for your reader. With complex datasets, often the best or only way of doing this is through graphical displays – graphs, maps, contingency tables, and so on – while for other data (typically from qualitative methods) this may be more effectively communicated through extended prose.

WAYS OF PRESENTING DATA

Data can be presented in many different ways. The best way of presenting your data will depend on the type collected and the focus of your dissertation and argument. Presenting data in written form works well for interview or focus group quotes or communicating small numbers (typically three or fewer) of variables – such as comparing two or three specific data points (Montello and Sutton, 2012). If you are trying to communicate four or more variables, using tables or graphs is typically more helpful in simplifying the content and making it easier for a reader while often using fewer words – as the following fictious examples illustrate:

Average temperature and rainfall recordings in New Delhi are 29.8°C high and 15.2°C low with 20 inches of rainfall in March, 36.7°C high and 21 decC low with 14 inches of rainfall in April, and 39.6°C high and 25.3°C low with 20 inches of rainfall in May (Climate Data, 2025). Compare this prose with the succinct presentation of data in (Table 8.1).

Data tables can be used to effectively depict quantitative data (and, at times, some forms of qualitative data) and communicate meaning and connections. Other visual formats – graphs, maps, and charts – can provide further opportunities to effectively communicate meaning and understanding, fostering insights into and understandings of key relationships, trends, outliers, and other features within your data. The rest of this chapter provides brief guidance on a number of possible ways of presenting your data – based on the core premise that data presentation is fundamentally about effective communication (for a more comprehensive discussion on data presentation, see Pering, 2026).

Table 8.1 Example of Data Presentation in a Table: Average Temperatures and Overall Rainfall, New Delhi, March–May 2025

Month	*Average Maximum Temperature (°C)*	*Average Minimum Temperature (°C)*	*Overall Monthly Rainfall (in)*
March	29.8	15.2	20
April	36.7	21.0	14
May	39.6	25.3	20

Source: Climate Data (2025).

To ensure the effectiveness of these media it is vital to ensure that the features included in the image/graphical product – including any captions or other text – are clear and legible at the size you are presenting the image. Do not add unnecessary detail or 'decoration' to images that dilutes the key information being communicated. Do, though, think carefully about the colours used in graphs, maps, and charts – both in terms of conventions within the field (whether colours associated with political parties or land uses) and in terms of accessibility concerns and the ease with which the viewer can differentiate between shades or colours. And always check that your image rescales to the necessary size without undue pixelation, loss of quality, or with details being rendered too small to clearly read.

Effective communication of data through graphical media also requires two other key considerations. First, that the material is clearly linked to and integrated with the main text. This means ensuring the material is suitably placed and referenced in the text so your reader knows the purpose of the material, how you are using this in your overall narrative, and can readily connect data and text.

Second, ensure all tables, graphs, figures, and similar have a label or caption, and that these are included in the list of tables/figures in the front matter. Labels and captions should be logically numbered and contain basic descriptive information that explains the content being presented (e.g., Figure 1: Constituency boundary map from the 2023 Boundary Commission for England Review (source: Boundary Commission for England, 2023), or Table 1: Average monthly precipitation levels in South Africa, per province, 2014–2016 (derived from: South African Weather Service, 2025)) Clear labelling also extends to titles, variables, and units (as applicable), and to any other mark-ups required (for instance, if truncating axes). These details should be sufficient that the reader can understand the illustration as a stand-alone object (Field, 2016).

COMMUNICATING QUANTITATIVE DATA EFFECTIVELY

If you are using quantitative data this needs to be clearly presented through descriptive and/or analytical representations as appropriate (Field, 2016). The style of presentation depends on the type of data and analysis being used and purpose you want the data to serve. Basic exploratory and descriptive techniques, including graphs, tables, and plots, can be used to explore data and present key findings (specialist and analysis-specific data presentation guidance can be found in Chapter 11 in Montello and Sutton [2012] or see Pering [2026]).

To illustrate the frequency of values within different ranges, a stem and leaf diagram (see Figure 8.1) is a simple but effective approach.

Scatter plots can be used to show the relationships between two variables and are often generated during data analysis. These plots might also include a line of best fit and be used to support or illustrate analysis of data and correlation between variables (e.g., Figure 8.2).

Graphs can communicate data through pictorial representations, typically for general findings rather than specific patterns and to illustrate *distributions* of variables. Bar graphs (for nominal or ordinal variables – Figure 8.3), histograms (for discrete or continuous data on an interval scale – Figure 8.4), and line graphs (for continuous

0	4, 4, 7
1	3, 5, 6, 8
2	0, 0, 2, 3, 5, 6, 6, 9
3	1, 1, 3, 4, 4, 5, 5, 6, 7, 7, 7, 8, 9
4	0, 1, 1, 3, 4, 5, 5, 7, 8, 8, 9, 9
5	3, 3, 4, 5, 6
6	2, 2, 5, 6, 7, 7, 9
7	0, 4, 7, 8
8	2, 4
9	1

Figure 8.1 Illustrative Stem and Leaf Diagram.

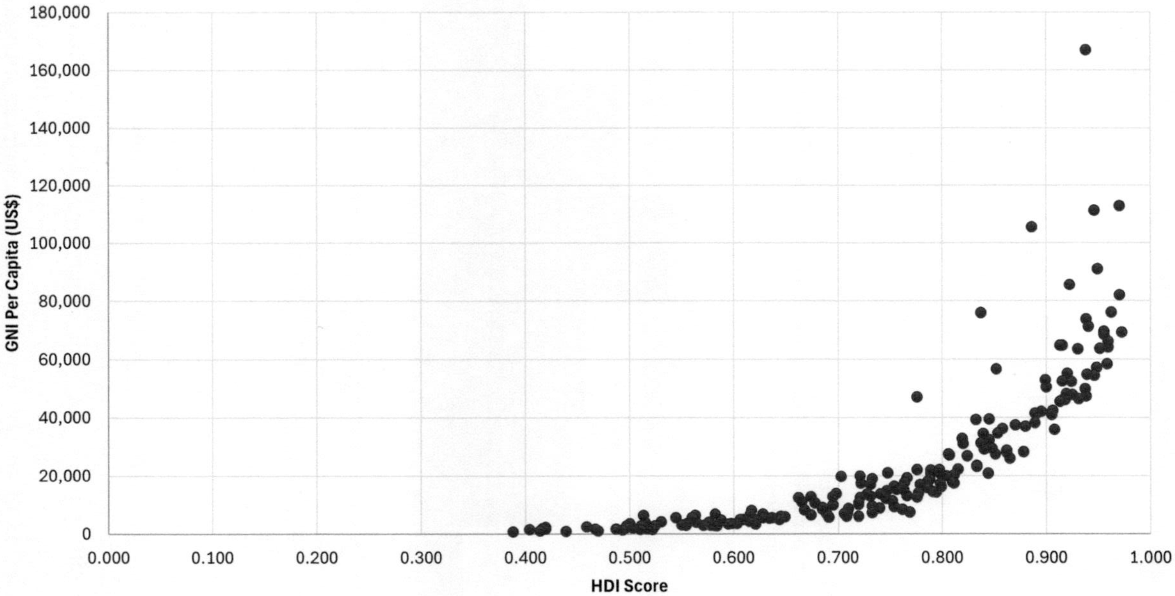

Figure 8.2 Illustrative Scatter Plot Using Data on Human Development Index (HDI) Scores and Gross National Income (GNI) by Country. (Based on data from UN, 2025).

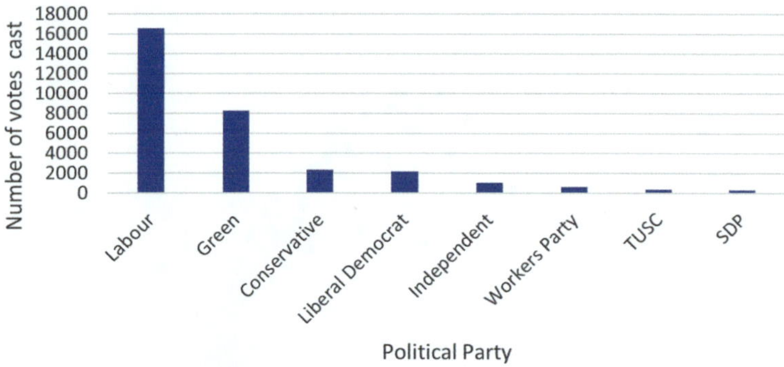

Figure 8.3 Illustrative Example of a Bar Graph Showing Number of Votes Cast per Political Party in the Sheffield Central Ward in the 2024 General Election.

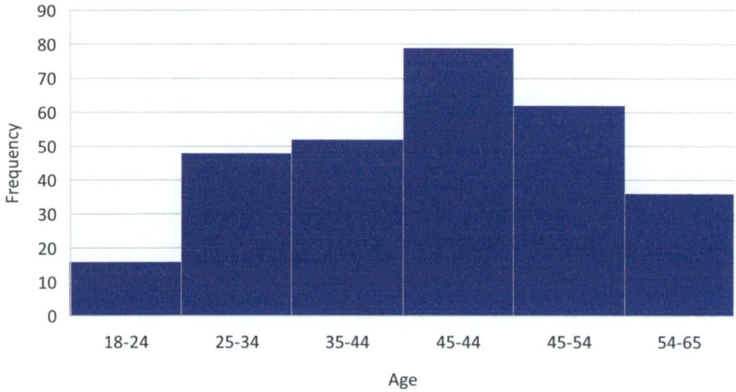

Figure 8.4 Illustrative Example of a Histogram Showing Attendees at an Event by Age.

variables – Figure 8.5) typically provide visual depictions of data using variables along the X-axis and frequency on the Y-axis. Circle diagrams – or pie charts – can be useful in presenting nominal data (Figure 8.6). As with all data, make sure that any graphs presented serve a specific purpose for your dissertation. Avoid the temptation to include extra materials simply to demonstrate the amount of data collected or analysis done. Rather than helping strengthen your work, including tangential or duplicate content undermines the clarity of your argument and uses up words unnecessarily.

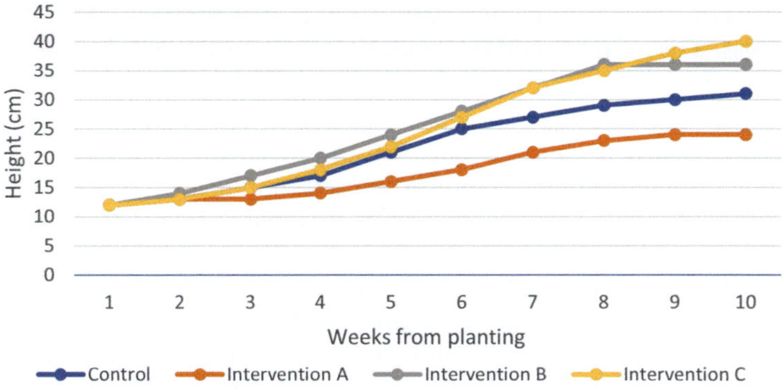

Figure 8.5 Illustrative Example of a Line Graph Showing Differences in Plant Height Over Time for Different Growth Interventions.

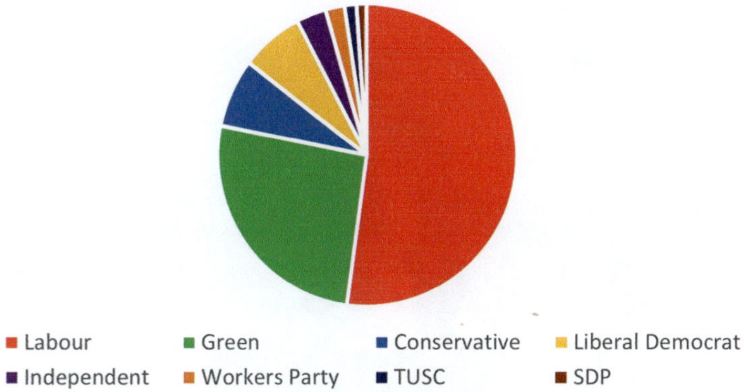

Figure 8.6 Illustrative Example of a Circle Diagram Showing of Votes Cast per Political Party in the Sheffield Central Ward in the 2024 General Election.

USING MAPS

Maps can be an invaluable means of communicating your data and data points, your case study location, or to highlight relevant natural or anthropogenic features for your work. You can also use maps to illustrate the location of transects or mobile methods, or to locate a series of images or data points within a site. You might also develop specialist maps that illustrate precipitation levels, voting intentions, vegetation cover, and myriad other factors over different spaces and scales. They can be used to depict concerns across multiple scales – from the body to the global (see, for instance, Rieger et al., 2022).

You might use reference maps to show earth surface features, whether terrain, infrastructure, urban development, and so on (as well as other features relevant to the scale and focus of your map). Thematic maps can be used to show the spatial distribution of a particular variable: these will not show extensive data about the earth surface but communicate geo-referenced data in an accessible visual manner – such as human development index scores or average natural vegetation cover (e.g., Figure 8.7). If working with time series data of a particular variable, mapping this over time can be a powerful means of communicating change. Montello and Sutton (2012)

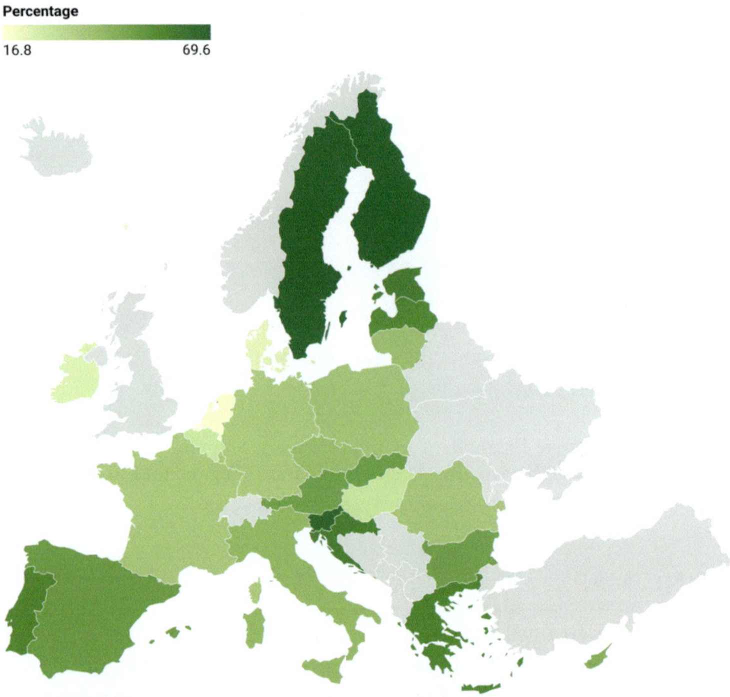

Figure 8.7 Illustrative Thematic Map Showing Average Natural Vegetation Cover by European Country (Derived from Eurostat, 2025).

provide useful further guidance and discussion on differing map projections and levels of generalisation.

You can also make use of symbols on maps to communicate data – this can include the common features of contour lines, symbols for specific features or buildings, the colour 'coding' used for rivers, roads, and the like. Symbols can be used to creatively communicate information in different ways: for instance, you might use a series of graduated circles to communicate differing sizes of urban settlement, numbers of homicides per city, and so on (see Figure 8.8).

Choropleth maps can be used to map quantitative variables that are measured by spatial area (iconic examples of these maps include colour-coded maps of election results in the UK or the United States). While these can be effective, they can give a misleading impression or sense of relative importance if you are working with population related variables as they cannot differentiate for

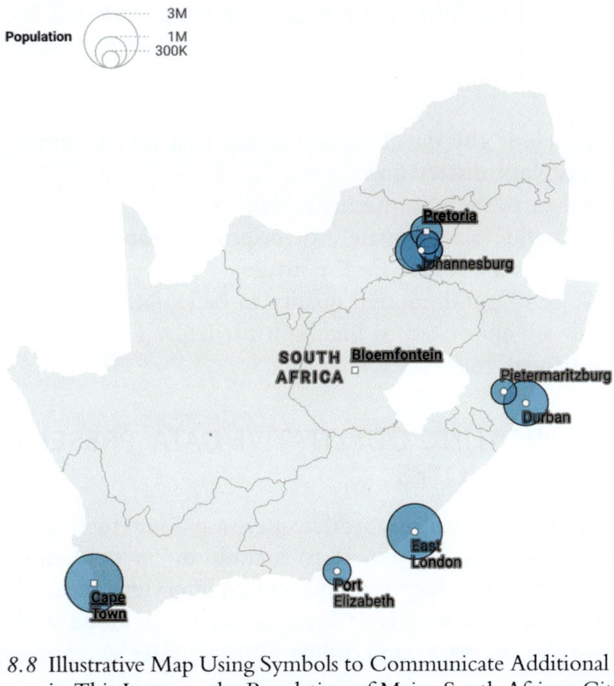

Figure 8.8 Illustrative Map Using Symbols to Communicate Additional Data – in This Instance, the Population of Major South African Cities.

population density. In these situations, a cartogram – which distorts geographical area by population – may provide a more useful and accurate representation of your data.

PHOTOS AND ILLUSTRATIONS

Photos and other illustrations can be highly effective in communicating information to your reader, but do not use these to simply 'decorate' your dissertation – they need to serve a purpose. If you have used photo elicitation, sketching, or another visual methods-based approach to your dissertation you should consider including a selection of such images to support and illustrate your discussion.

You might also want to use an image, or small number of images, to illustrate the context of your field site: for instance, in a dissertation on public perceptions of coastal erosion management policies in Norfolk you might include some images of the specific stretch of coastline. While some images may be integral to the analysis, others might be used to illustrate the context of your work and be better included in the background or methods section of your work.

If you do include images, ensure that

1. you include the details of the image in a list of figures at the start of your dissertation;
2. you refer to them in the main text;
3. you provide an adequately descriptive caption for the image (including the source for the image);
4. they are of good enough quality to be legible and, where relevant, are annotated to highlight particular features of interest or concern.

COMMUNICATING QUALITATIVE DATA: PRESENTING INTERVIEW QUOTES

If interviews or focus group discussions are a key method in your dissertation you should expect to include direct quotations from these in your results chapter. In presenting interview quotes, shorter excerpts of up to a couple of lines would be included within the main body of the paragraph, while longer quotes would be offset.

For instance, "this is a short quote embedded in the text" (George, 11/3/2025), and

> this is a second quote which continues ad nauseum about how to write a good dissertation. In making the argument about how to present quotes, the author really goes on and on and belabours the point. But at least can now see the different style of presenting quotes.
>
> (George, 11/3/2025)

As with these examples, you need to include appropriate attribution for all quotes, whether using standard citation practices for archival or media content or attributing an interview quote to a respondent. You must also clearly identify where you are quoting interview responses through use of "" or ' ' (depending on your departmental style guidelines). It is advisable to avoid automatically emphasising quotes using italics or other font changes. If there is a particular part of a quote you want to draw the readers' attention, italicise that component only (and include 'emphasis added' in the attribution), or if the respondent placed particular emphasis on a word/words then you should italicise this content specifically to draw attention to the importance of that particular wording. This communicates to the reader that you are drawing attention to this for a reason, which you should develop in your analysis of this material.

As with other forms of data presentation, do not simply include quotes for the sake of it. Each quote should add to your narrative. If you have multiple interviewees making the same point, be selective and use one or two key quotes in the accompanying analysis to indicate that this was a wider point of concern or comment – for instance, "Reflecting comments from multiple interviewees, Margaret (17/8/2024) argued" or "As with many residents interviewed, Mohammed (14/10/2023) noted".

Where you have differing opinions between interviews you can place the differing views into conversation in your writing and use these to develop a critical discussion and analyse these differences. In so doing, build your argument – which does not necessarily mean agreeing with one viewpoint or another, but rather teasing out what this difference tells you about the issue you are researching.

ATTRIBUTING INTERVIEW (AND SIMILAR) QUOTES

How do you best attribute interview and similar quotes to help the reader follow your thinking and data? If you have ethics approval for and agreement from individuals to include their names, then feel free to do so. More commonly, you need to ensure the anonymity of participants. In this case, avoid talking in overly general terms about 'a participant' as this does not help the reader understand if different quotes or discussion relate to the same person or to different people. You might opt for a very basic differentiation (such as Participants A, B, and C). While this provides some indication of whether the same participant is being talked about or quoted in different places, it can be difficult to follow and fails to provide any indication of a respondent's positionality. Most supervisors would instead encourage you to make use of false names – pseudonyms – for your respondents. This approach generally makes it easier for the reader to keep track of different respondents and careful choice of pseudonym can give some impression of the respondent (e.g., gender). Even better is to use an appropriate pseudonym accompanied by some very basic contextual detail. For instance, "Matilda, an established business owner in downtown Lagos" or "Xolani, an undergraduate university student". This helps the reader both identify respondents through the dissertation and can be useful in supporting your analysis.

Remember also to include the date of the interview in the attribution. Adding this information is important if you are developing a narrative based on multiple consecutive interviews to chart changes over time or if there was a significant event during your fieldwork period that may have resulted in a change in responses.

If you have conducted focus group discussions you can include standalone lines or responses from individuals as well as longer excerpts of interactions within the group. Ideally these will be presented, as with interview quotes, with attribution to the relevant (anonymised) individual(s). Again, it can help to nuance the discussion if you can trace specific respondents and their input through the discussion – this can help you and the reader understand if there are a small number of very prominent or dominant voices, or if there is widespread agreement on a topic, and so on. However, given the nature of focus groups it is not always feasible to reliably identify and attribute specific quotes to individuals. In which case, you might need to present such quotes simply as "Focus group participant, Date".

If you have kept a research diary and want to quote this document then the same core principles apply as with citing published sources, using a citation of (Research Diary, DATE OF DIARY ENTRY). However, you do not include your research diary in your reference list as it is not publicly available. You might find it useful to quote your research diary if you want to provide a description of a place, event, or encounter (an excellent example of this is Elijah Anderson's (2004) thick description of Reading Terminal Market in Philadelphia, USA), to illustrate and discuss an encounter in your research, to provide contextual detail, or recount an interaction during your research.

It is generally accepted that interview quotes in geography dissertations can be lightly edited to 'tidy up' and remove clear redundancies such as 'umms', 'ahhhs', and 'likes' as long as these do not change meaning or emphasis of what was originally said. Similarly, you can 'abridge' quotes as long as this does not change the meaning nor provide a deceptive interpretation on the original materials. If you do abridge a quote this should be clearly denoted in the quote using a series of three dots (ellipsis) to demarcate the section(s) that have been removed from the quote. For instance, "Since volunteering with the river restoration group that feeling, you know, my overall sense of belonging has developed" may be abridged to "Since volunteering… my overall sense of belonging has developed". The key principle here is that while you might edit a quote for clarity and succinctness, you must not knowingly change the meaning or emphasis of the response, alter the substantive statement (e.g., by removing a 'not' to invert the statement made), nor remove substantive material of relevance.

CHAPTER SUMMARY

Geography dissertations may be based upon a vast array of data and analysis, from analysis of archival sources to remote sensing and GIS, from measuring volcanic gases to auto-photography, from interviews to census data. There is, therefore, no single 'correct' way of presenting data. Instead, the most appropriate and effective way of presenting your data needs to be determined on an individual basis. This chapter has outlined foundational approaches to presenting common types of data in geography dissertations. Regardless of the approach adopted it is imperative that the presentation is clear, accurate, and accessible. Key to this is to be selective – not in terms of cherry-picking data to fit your narrative but in ensuring you present sufficient data to underpin your argument but without falling into the trap of unnecessarily including duplicate or tangential materials simply to demonstrate how much work you have done. Once you have effectively presented your data and related analysis and discussion, the logic and connections to your conclusions should be readily apparent.

9

STEP BY STEP

Conclusions

The concluding chapter is the last substantive part of your dissertation. This chapter provides a succinct, powerful closing summary of your main argument and findings, and makes clear connections back to the literature and articulates how your work builds on this existing understanding and knowledge. A poor conclusion will be disconnected from the framing literature and/or your results and findings, may introduce new evidence or debates and ideas, and sow confusion and uncertainty as to whether you really understand what your dissertation was about. This *step-by-step* chapter provides an overview and summary of the purpose of the concluding chapter of your dissertation and guidance on how to ensure this effectively delivers against key expectations and common marking criteria.

WHAT DOES A CONCLUDING CHAPTER DO?

Fundamentally, the concluding chapter of your dissertation brings your argument to an end: it draws together the strands of evidence and discussion to reiterate your key claims and findings. You want this to be sharp, clear, and concise. It is not a space for conjecture, nor to be introducing new literature, ideas, or evidence. Rather, it should leave the reader with a powerful final message. To do this, your conclusion should

1 revisit and summarise the main ideas and themes to ensure your audience understands what your research is about (i.e., very briefly summarise the main focus and purpose of your work);

DOI: 10.4324/9781003584315-9

2 clearly reprise your main argument – and in so doing, relate back to the territory or field of your research and niche you are addressing, articulating why your work matters and how it can contribute to discussions within the (sub)discipline;
3 building from (ii), you should take this opportunity to identify how your thinking could be taken forwards – specifically how further research could develop your insights and take these forwards. This is an opportunity to demonstrate that you understand your field of work sufficiently to identify how other work could further develop your insights.

STRUCTURING YOUR CONCLUSION

While your introduction is likely to be structured to move from the big picture down to the specific focus of your research, your conclusions will do the opposite. You need to first communicate *your* key findings and argument based on the specific focus of your research, and then highlight their wider implications by articulating the originality of your research and what it offers as a new insight or evidence for a concept, a theory, a method, or similar.

In practical terms, your conclusions should start by summarising and reiterating your findings and how they address your research questions/hypotheses. You should then link these to broader debates and issues, explaining why your work matters and how it contributes to the discipline. In some instances this core section is most effectively structured by research question, allowing you to highlight how you have addressed each in turn. However, this is not always the case and your decision in this regard will likely be framed by whether you have structured previous sections (e.g., results and discussion) by research question.

If you have adopted a more overarching or narrative structure to previous sections, it may well be more effective to continue with this approach in your conclusions. Whether explicitly structured by research questions/aims or not, your conclusions do need to clearly identify how you have addressed these. This is also, therefore, a crucial moment for you to assess whether your dissertation has done what it claims and to sense-check your work overall. This process can help you to ensure that your red thread is clear and consistent across your dissertation and ensure your conclusions draw this to a close. It is the final opportunity to check that your title fits your findings, that you have addressed your research questions, and that your main findings and argument are reflected in your conclusions. If you realise this is not the case, then this is moment to pause and consider how best to adapt your dissertation to ensure consistency throughout and to ensure your dissertation is a clear, coherent, and compelling piece of work.

This process can also help you to identify ideas for the very final part of your conclusions – your recommendations for further avenues for research that arise from your project. This is not intended as a moment for you to cast doubt on the quality of your work. Rather, this is key opportunity to critically reflect on the research process and recognise that while you have contributed to knowledge, the potential remains for other researchers to further build on this work. Being able to identify what this potential might be is a good indicator of your critical engagement with the research journey.

WRITING YOUR CONCLUSION

Writing a full draft of your conclusions can only realistically be done once you have drafted most, if not all, of the main body of your dissertation. However, you can do a lot of preparatory work in the meantime. While writing other sections of your dissertation you can develop bullet points and short notes for your conclusions that encapsulate key ideas and framing from existing literature, key findings and argument(s), and thoughts on what else you would have done with more resources and time. This note-making process can help ensure your conclusions capture the key ideas and thinking from across your dissertation, while also significantly speeding up the process of writing this final chapter as you already have significant material to work with.

While you wouldn't want to write the conclusions in such a formulaic way, it can be useful to plan this section by asking yourself the following questions – and then using your answers to help develop your chapter:

1. What is/are the main finding(s) of my dissertation?
2. What is the 'contribution to knowledge' of my dissertation? (In other words, what am I saying here that is new or different to existing literature?)

3 Which key fields of literature or ideas have informed my work and what can these (and other areas of study) learn from my work?
4 If I were able to do more work on this topic, what would I do and why? What would the extra work allow me to do and learn?

This process also allows you to review your dissertation and ensure that it does what it claims to do and that there is a clear and robust argument to your work. You should use this moment to ask some other questions of your work overall and potentially go back and revise other sections of your dissertation.

1 Does the title clearly and accurately reflect the content and focus of the dissertation? If not, it needs changing.
2 Have I answered my key research questions in this dissertation? If not, you need to revisit your analysis and focus *or* revise your research questions.
3 Have I made it clear how I have built on and added to existing literature? Is it clear what is 'new' in the dissertation, what the contribution to knowledge is? If this is unclear or incomplete then revisit the relevant sections of your dissertation and emphasise and clarify these connections and contributions.
4 Have I made effective use of (sub)headings to clearly scaffold my content and signpost for the reader what material is in each section? Do these give a sense of the logic and flow of the work, and are they all relevant to the 'red thread'? If necessary consider adding new/additional sub--headings to help strengthen the scaffolding and/or revisit existing sub-headings and edit to ensure they clearly signpost the structure of the work for your reader.
5 Do each of my other chapters have their own concluding sections that (i) summarise the content of that chapter and the key point/argument made and (ii) make clear and logical links to the following chapter? Do these chapter-level conclusions feed into/resonate with the overall conclusion chapter? If not, then take the opportunity to remedy this.

POTENTIAL FOR FURTHER RESEARCH

A key, but oft neglected, part of the concluding chapter is to think about how further research could develop your ideas further. What would be the next step(s) in the incremental accrual of knowledge on this topic? By identifying these next steps you are able to indicate your critical engagement with the research process and indicate your ability to think critically about data, findings, and further contributions to debates. While there are often some obvious ideas that can be suggested – more time, more case study sites, more participants – it is important to think beyond these. What have been some of the threads of analysis you haven't been able to follow? Have there been some linked but tangential considerations that could be explored? Would the use of a particular additional dataset allow a more robust examination of a particular sub-theme or question?

Setting out what one or two further areas of research could be and why these questions or ideas matter is key. Ideally you want to locate these as being a continuation of or as building on your work while being clear that these possibilities are beyond the scope of your dissertation. The following examples illustrate some effective ways of positioning these ideas:

> This dissertation focusses solely on the socio-environmental impacts of changing flood risk in Lympsham on the Somerset Levels. Further research based on this dissertation to explore concerns across other towns on the Somerset Levels would do Y.

> Despite significant efforts to ensure diversity in interview respondents (see methods chapter), the limited number of older (65 years plus) respondents has resulted in a primary focus on intergenerational differences between younger cohorts. Further research on this topic focussing on older demographics would allow Y.

Crucially, you need to ensure you are not overly negative in your conclusions – so think carefully about where to locate discussion of limitations (methods, conclusions, etc.) and how to balance acknowledging these limitations and how they have affected your findings with presenting these as an opportunity for someone else to further develop insights into the topic.

CHAPTER SUMMARY

Your concluding chapter needs to clearly and powerfully tie the thread(s) of your dissertation together and make a compelling closing statement. Typically, this is a relatively short chapter in a dissertation – you are not introducing new ideas, concepts, or data, but instead wrapping-up the preceding materials to reiterate your main argument. As with previous chapters discussed, it is important not to overclaim the importance or significance of your work – but you do need to be clear and confident in providing an evidence-based assertion of your contribution to knowledge. It is worth remembering that your conclusions will be the last substantive part of your dissertation that your reader engages with – you want to ensure their final impression of your work is a positive one.

10

STEP BY STEP

Introduction

The introduction can be one of the hardest parts of your dissertation to write. It is also one of the most important chapters of your dissertation, setting the tone for the rest of the document and providing an early impression of the quality of your research. Your reader should finish your introduction with a clear sense of the overall rationale for and importance of your topic, the key argument of your dissertation, its structure and flow, and the core debates you engage with. This *step-by-step* chapter explains what make a good introductory chapter and then guides you through the key content and structure of this content.

Perhaps counterintuitively, your introduction should be one of the final chapters you write. For your introduction to be effective you need to be confident not only about the flow and argument of your dissertation but also in explaining in a clear and succinct manner what your work is about, why it matters, what your argument is, and how your dissertation is structured to effectively make this argument. Once you have the rest of your dissertation written – or at least drafted – this becomes much easier to do, and ensure that your introduction makes a compelling case to your reader that your dissertation is worth reading.

WHAT MAKES A GOOD INTRODUCTION?

A good introduction will do five main things:

1 Provide the broad context or big picture that frames your dissertation.
2 Set out the specific focus of your dissertation.

DOI: 10.4324/9781003584315-10

3. Get your reader's attention and make them want to read your work.
4. Indicate the key findings and main argument of your dissertation.
5. Explain to the reader the flow and structure of the rest of your dissertation.

An alternative way of thinking about this is to conceptualise your introduction as answering five key questions:

1. What is the dissertation about?
2. Why does it matter?
3. How does this dissertation relate and contribute to existing literature?
4. What is the key argument and/or findings from the dissertation?
5. What is the structure and flow of the dissertation?

These questions and prompts are focussed on helping you to demonstrate why your dissertation is important and to provide an overview of what is to follow. If you are unable to assertively respond to these questions you should take time to revisit your work and clarify these key features. You need your introduction to be confidently and competently written – you want your reader to finish this material already confident in the quality of your work. If your reader gains a poor impression from your introduction – of the rigour and robustness of your project, of the importance and relevance of the topic, of the conceptual or theoretical basis for the work, etc. – this negativity can inform their reading of your work overall. It is therefore in your own self-interest to ensure your marker gains a positive impression from this opening chapter (see also Chapter 14 – *Remember your audience: Keeping your marker onside*). To help with this, it can be effective to provide an opening 'hook' for your reader that grabs their attention and signals the urgency and importance of your work.

HOW TO START YOUR INTRODUCTION

Your opening statements or sentences are key to gaining the reader's attention. This is often achieved through conveying a particularly powerful issue or moment related to or emerging from your specific focus, or from establishing the urgency and importance of the broader issue framing your research. There are a number of commonly used approaches for this, including the following:

1 A provocative or otherwise engaging direct quotation from an interview, an article, a news story, or your research diary.
2 A paradox or question.
3 A link to a high-profile contemporary issue, event, or news story.
4 A short anecdote or observation from your research.

Various of these approaches are illustrated in the following examples that demonstrate how creative and engaging ways to 'hook' the attention of your reader at the start of a dissertation can work (for further details on using hooks to catch your reader's attention, see Chapter 1 – *Thinking about your dissertation*). Each 'hook'-based opening is paired with a less engaging alternative opening for comparison

For a dissertation on the impacts of climate change on flooding in the UK:

> Global temperatures have already risen by at least 1.1°C since the late 19th Century (IPCC, 2021) and are forecast to continue rising in the coming decades leading to increasing risks of major flooding events.

OR

> This dissertation analyses how climate change may have affected flooding in the UK in recent years.

For a dissertation on urban governance in Kenya:

> 4.4 billion people – 56% of the world's population – live in cities, with these figures expected to increase over the coming years (World Bank, 2023). In Kenya, as across the African continent, urbanisation and urban growth are proceeding apace, leading to major challenges to urban governance and infrastructure.

OR

> African cities are growing quickly meaning there are various urban governance challenges.

For a health geography dissertation on rural healthcare in South Africa:

> I arrived at the hospital early this morning. One of the medical staff showed me to a spare room in a temporary housing facility made from what looks a portacabin. It is clear already that the wards are overcrowded and doctors overworked: people are queuing in the shade of trees for most of the day outside dilapidated buildings, with clinical waste dumped nearby. The few consulting rooms that are here are small and dark, the air close and fetid. (Research Diary) This quote from my research diary highlights the pressures facing rural health care provision on a day-to-day basis in South Africa.

OR

> Rural health care in South Africa is of inadequate quality and many residents struggle to access adequate service provision, often in poorly maintained facilities.

For a dissertation on rewilding practices in the UK:

> Dartmoor wild boar sightings prompt suspicions of guerrilla rewilding.
>
> (Horton, 22 February 2025)

> Illegal release of four lynx 'could be start of rogue rewilding campaign.'
>
> (Vevers, 15 January 2025)

> Rewilding – the process of restoring an area of land to its natural state, often through the reintroduction of species of wild animals – is increasingly prominent in public debates on conservation in the UK. While recent headlines have focussed on illegal ('guerilla') rewilding, this dissertation focuses on the impacts of a permitted release of beavers on water quality and biodiversity at one site in Scotland.

OR

> Focussed on one site in Scotland, this dissertation considers how the release of beavers might have an impact on biodiversity and water quality.

As these examples should illustrate, there is no single 'correct' way to being your introduction. Granted, the alternative sentences are deliberately developed to provide contrast, but these hopefully highlight how an engaging opening can make a substantively positive impression compared to a bland, generic statement.

Once you have set the groundwork with this opening paragraph, the rest of your introduction needs to strike a balance in providing sufficient background or foundational information for the reader in explaining key concepts and locating your work within the relevant historical and/or geographical context and conceptual/theoretical framework. The communication of these different components informs the structure of the rest of your introductory chapter.

STRUCTURING YOUR INTRODUCTION

One of the easiest ways to envision your introduction can be as an inverted triangle (Figure 10.1) that begins by setting out the broad context for and importance of your project, with subsequent paragraphs and sub-sections progressively narrowing the focus down to your specific topic or research problem and key argument. This approach commonly uses the Create a Research Space (CaRS) model (Swales, 1990) in setting out the structure of your dissertation.

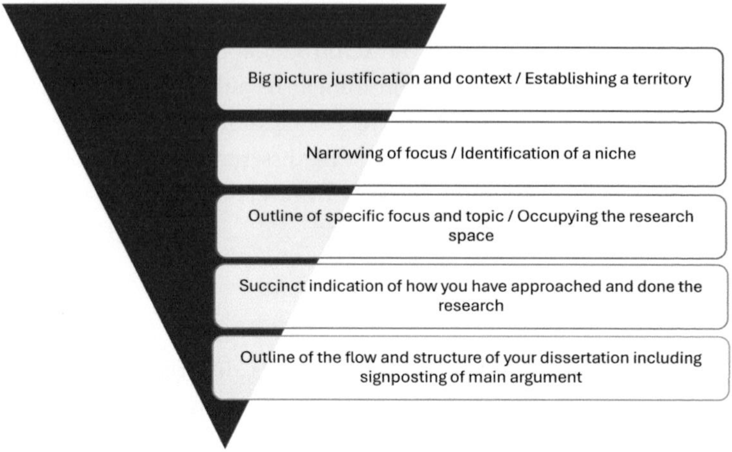

Figure 10.1 An Inverted Triangle Structure for Introduction Adapted with the CaRS Model.

THE CREATE A RESEARCH SPACE MODEL

The CaRS model provides a series of three key steps to help develop a clear and powerful introduction:

1 Establish a territory
2 Identify a niche
3 Occupy the research space

Establish a territory

This is where you set out a claim to the importance or *centrality* of your research and locate this within current debates. By succinctly setting out the broader context that frames your work you signal your awareness of existing relevant literature *and* your understanding of how your research relates to these debates. This should not be a lengthy review of the literature (this is done in Chapter 5 – *Step by step: Literature review*), but a brief precis of key issues, ideas, or debates that provides the broad rationale for your dissertation. This can include introductions to key debates and concepts, summaries of pertinent conceptual or theoretical materials, and some contextual background that frames your work.

To help narrate this territorial claim common rhetorical devices may help with concisely introducing or segueing into relevant literature, providing a sense of purpose or direction to the narrative. Potentially useful phrases include 'Current concerns with', 'Recent debates have', or 'Emerging research has begun to identify' – as illustrated in the following example:

> Concerns with the increasing intensity and frequency of extreme climatic events linked to climate change have focussed attention on the suitability of urban environments to cope with flooding and other hazards.

Identify a niche

Once you have established the broad territory of your research, you need to identify where your research fits within these debates and the particular need, gap, or niche it addresses. Translating this into dissertation marking criteria, this is where you set out what your original contribution to knowledge is. To do this you begin by identifying what is known and the limitation(s) to this (in other words, what is not yet known). Having identified this absence, you present your work as responding to or addressing this gap in knowledge – remembering that you are doing so at a very limited scale and with a specific focus appropriate to a dissertation.

Effectively identifying the niche and signposting your response requires a succinct demonstration of your critical engagement with existing literature to highlight relevant strengths and weaknesses or oversights. Be careful to avoid hyperbole and overly generalised or sweeping claims in making your case for a gap in existing understanding: declarative statements such as 'no research exists' are easy targets for markers to critique and rebuff.

For instance, if your dissertation explored how African cartoonists engage with global politics you should avoid declarative statements such as "Despite political geographers' extensive engagements with popular geopolitics, there is no research on African popular geopolitics". Even if you haven't found these materials, such research does (in this case) exist. A more judicious wording such as "Despite political geographers' extensive engagements with popular geopolitics, there remains relatively limited attention on African popular geopolitics". This phrasing still demonstrates your critical engagement with existing literature and highlights a gap that the work responds to but removes the easy critique for your marker.

Remember that you are identifying a *niche* to address, you are not trying to change an entire field of study nor to propose an epoch-defining concept. Your aim here is to make a specific and identifiable yet minor, incremental addition to knowledge or understanding. Articulating the gap you are responding to can be done using phrases such as "Within these broader debates there remains limited engagement with", "The opportunity remains to advance knowledge through consideration of", or "However, existing research is lacking critical understanding of/perspectives on". Clearly there are many ways in which you can communicate the gap you are writing to and your contribution to knowledge – the key concern here is that you know what the gap is and how you are responding to it (and that you communicate this to your reader).

Occupy the research space

After you have established the territory of your work and identified the niche you are addressing, you can set out how your work responds to this. This is where you outline what your dissertation does, its main purpose, and to signpost the key argument and findings being mobilised. Frequently used phrases here include "This dissertation argues", "This work evidences", "Drawing on detailed empirical evidence, this dissertation argues", "The main objective of this dissertation is", or "The main objectives of this dissertation are two-fold". To support and legitimise these claims you should briefly outline the main findings. It can be useful, although not always necessary, to signal here also how you reached these conclusions – for instance,

> The first objective of this dissertation, to investigate long-term precipitation and mean temperature change in the Yeo catchment area, was realised though analysis of HadUK-Grid data (at 1 km^2 resolution) of relevant variables. Over the 70-year period under investigation, changes in precipitation and mean temperature are identified as….

Signposting the key methods and findings here succinctly is appropriate, with further detail on the data collection methods, data analysis techniques and technologies, and further detail and discussion on findings following in the relevant later chapters.

What you are trying to do here it to set out why your work matters and what the main findings are: in other words, how you are claiming to occupy this research space. Here you should be clear in communicating what your key argument is and how the rest of your dissertation builds to this conclusion. This is where you substantively start the 'red thread' that runs throughout your dissertation (see Chapter 4 – *Good writing*).

PRESENTING YOUR RESEARCH AIMS AND QUESTIONS

One of the most common questions asked in relation to the introduction for a dissertation is 'Should I put my research questions here?' Unfortunately, there is no clear-cut definitive answer to this. As outlined earlier in this book there are four typical locations for you to include your research questions (in the introduction, as a short standalone chapter, in the literature review chapter, or in the methods section). Regardless of where you include your aims and questions or hypothesis/hypotheses your introduction needs to provide a clear signposting of the key focus and objectives addressed in your dissertation, and findings developed in relation to these. Broader advice on how to present your research questions is provided in Chapter 2 – *Supervisors, support, and common questions*.

SIGNPOSTING YOUR STRUCTURE

Once you have set out the focus and contribution of your dissertation, you should outline how you are going to tell this story. This means outlining the structure of your dissertation and providing an initial guide as to the focus of and flow between the chapters. This section should be relatively short and include headline details of the focus or argument of sections. Think of this final section as a roadmap of your dissertation, outlining the readers' journey through your work along with brief descriptive commentary – such that you both outline that 'section X follows section Y' and communicate very brief how each section builds your core narrative.

There are numerous ways to start this final section of your introduction, with commonly used phrases, including "This dissertation proceeds", "This dissertation is structured as follows. Chapter two", "Following this introduction, chapter two", or "The next chapter". As well as signposting the flow and structure of your dissertation, this outline overview should illustrate that your dissertation follows a logical structure and has a clear argument that each chapter contributes to. At the following illustrate adding brief details on the focus and contribution of each chapter provides a much more effective and compelling outline of a dissertation that a perfunctory list of chapter titles.

> **Example 1:** "This dissertation is structured as follows. Chapter two reviews existing research on coastal microplastic pollution. Chapter three outlines the data collection and analysis methods used, with results reported in chapter four. Chapter five presents the discussion and analysis of these data, before chapter six offers conclusions on microplastic pollution in Carmarthen Bay."

> **Example 2:** "In order to understand the everyday peace practices in Lagos this dissertation proceeds initially through a discussion of the intersection of peace studies and citizenship studies literatures, highlighting the role of formal and informal practices of governmentality in peace-building and -keeping endeavours. This is followed in chapter three by an outline of the participatory methods used with communities to develop insights into their everyday engagements with and practices relating to peace. Analysis and discussion of these data follow in chapter four, highlighting key emerging themes relating to intersecting scales of both formal and informal strategies and practices of governmentality and peace-building, and the contested role of informal, community-based policing/security organisations. This analysis builds towards the concluding argument (Chapter 5) of the need to focus on the 'everyday' in peacebuilding and for critical attention to be paid to the increasingly prominent role of informal actors in everyday urban governance".

While the second of these examples is considerably longer (145 words compared to 52 words for the first), there is a much clearer signposting of the rationale for and connections between literatures, methods, and findings. This extra detail illustrates the dissertation's red-thread and highlights the connections within and between chapters, your command of the materials and confidence in using these, and the robustness and reliability of your conclusions.

ENSURE YOUR INTRODUCTION AND CONCLUSION MATCH

Your introduction sets up the focus of your dissertation and sign-posts the key findings. Your conclusion brings together the key debates and reiterates the main findings of your research. These two chapters are the bookends of your dissertation – and it is vital that they complement each other and is obvious when they don't. To avoid this, the (final) draft of your introduction should be one of the last things you write so you can be sure it sets out what you *have* done rather than what you *planned* to do.

A useful activity to ensure the introduction and conclusion do correlate with each other is to read (or ask someone else to read) these two sections directly after each other (i.e., don't read any of the other chapters that come between them) and check that your conclusions deliver on what you say you are doing in your introduction. If they do, then all is well. If not, you need to revise your introduction (or conclusions) to ensure they accurately reflects the actual content and argument of your dissertation.

CHAPTER SUMMARY

The introduction for your dissertation does the groundwork not only for the focus of your project but in framing early impressions of the quality of your work. Alongside the abstract and other front matter (see Chapter 11 – *Step by step: Front matter*), the introduction plays a key role in expectation setting – expectations in relation to the scope and remit of the project, of the clarity of focus and writing, and of the general tone of the work that follows. As outlined in this chapter, careful planning and creative thinking will ensure not only that you not only create and claim a clear research space for your dissertation but you can convey to your reader the importance of this space. Getting your reader to invest in your work is key and the use of an effective 'hook' can be one of the most effective ways of ensuring this – with the hope being that this investment helps to keep your reader (your marker) onside throughout the journey (see Chapter 14 – *Remember your audience: Keeping your marker onside*).

11

STEP BY STEP

Front matter

First impressions count. The front matter of your dissertation is your first opportunity to make a positive impression on your marker. While this content is largely procedural, it is nonetheless important to take time to complete the sections carefully.

Typically, the front matter includes a cover (or submission) sheet, title page, declaration page (that the work is your own), acknowledgements, abstract, list of contents, list of tables, list of figures, and a list of acronyms. Some components will require you to populate a pre-existing template with relevant details, while for others you can use functions within your word processing software to automatically complete these. A few, though, will require more substantive engagement. Regardless, the quality and professionalism with which these materials are presented are important as they can inform your reader's expectations of the rest of your dissertation. Poorly presented or incomplete front matter suggests to your reader that the work overall will be rushed and weak, leaving you an uphill battle to convince your marker otherwise. This *step-by-step* chapter provides a guide to the common components of a geography dissertation's front matter, providing advice on how to effectively convey the required information while making a positive impression.

TITLE PAGE

The main purpose of this page is to communicate your dissertation's title and other key required information such as student number and/or name, word count, supervisor name(s), date of submission,

DOI: 10.4324/9781003584315-11

"Selling our soul": Community responses to the tourist-ification of Leith Docks, Edinburgh

Daniel Hammett

BA (Hons) Geography

GEO345

Student number: 123456479

Supervisor: Dr Jackson

Word count: 9,896

Figure 11.1 Example Cover Page.

and course/module title and code (see Figure 11.1 for a mock-up of a dissertation cover page). If your department has a template or guidelines for the title page then use this. Inputting the required information should be a simple process but take a moment to ensure you have an effective and accurate dissertation title.

WHAT MAKES A GOOD DISSERTATION TITLE?

Your title is the entry point to your dissertation. It is the first thing your reader will look at, so ensure it clearly and accurately reflects your research. A good title will communicate the main focus of the research, providing sufficient detail to signpost the content that follows but without becoming so detailed as to be off-putting. To help with this, many titles use a colon to introduce a subsidiary clause to the title, allowing you to indicate the big picture topic area *and* specific focus of your work. Other possible strategies can include using a short quote from an interview or other data as a 'hook' for the reader or using a question within the title (see Chapter 1 – *Thinking about your dissertation* for more detail). As the following titles for a notional project on immigrant investor programmes (a.k.a. Golden Passports) in the Caribbean demonstrate, it is worth getting your title 'right'.

Title 1: Citizenship by Investment Programmes
Title 2: A comparative and conjunctural analysis of the impacts of economic citizenship programmes in four small island developing states, with particular emphasis on real estate markets and public space
Title 3: Problematising the Role Citizenship by Investment Programmes in Caribbean Island States
Title 4: Global citizens, local impacts: Everyday encounters within economic citizenship programmes in the Caribbean
Title 5: Citizens of nowhere? How economic citizenship programmes undermine national identity in St Kitts and Nevis
Title 6: "You can't sell my identity": Local resistance to economic citizenship programmes in St Lucia

The first example is too broad and vague, leaving the reader with no sense of the specific focus or purpose of the work. The second is too complicated and detailed, meaning it holds little appeal to the reader (remember – keep it simple and straightforward [see Chapter 4 – *Good Writing*]). The third is better – it sets out a topical and geographical focus and signals a critical analysis and engagement (rather than a descriptive summary). The fourth example is more engaging, providing a clear and succinct sense of the focus of the work while hinting towards the broader literatures and contributions. The fifth option effectively uses a (rhetorical) question to engage ('hook') the reader into the discussion, while the sixth demonstrates how a short, provocative quote can be used to hook the readers' attention and signpost the focus of the work.

As these examples illustrate, a good title provides a clear, succinct signposting of the key themes and focus of the work. While a catchy or creative hook (a provocative quote from an interview, a pun or play on a popular culture reference, etc.) can be useful, the most crucial consideration is whether your title reflects what your dissertation does. In a variation of the classic adage of 'Read The Question, Answer The Question', a useful sense check is to 'Read the Title, [and check your dissertation] "Answers" the Title'.

To help ensure this, the title of your submitted dissertation should be one of the last things to finalise. It is, nonetheless, useful to have a working title from the start of the writing process: as a temporary 'place holder' this can frame your initial writing while recognising that this may evolve in parallel with your analysis and writing. This approach minimises the potential for time being wasted early in the writing process in agonising over the 'perfect' title which may not fit the final write-up or for you to feel compelled to write to fit a title that does not fit your results.

DECLARATION PAGE

For some departments, the 'declaration of originality' – the statement confirming that the work is your own and that you have appropriately acknowledged the contributions of others where appropriate – may be included on the title page. In other departments, this will be a separate page. Many departments provide a template phrasing that you have to include and 'sign' (electronically) to confirm that your work is not plagiarised, that you have not used or have fully declared any use of generative AI, and that the research design and data collection and analysis have all been done by you – unless otherwise declared and explained.

For most students there is 'nothing to declare' but if your dissertation is part of a larger project linked to a laboratory team or supervisor-led project it is essential that this is acknowledged and clarity provided as to what you have done and what contributions others have made. This statement also means you are confirming that you have not used generative-AI (such as ChatGPT, DALL-E, Gemini, copilot, and LLaMA) or other AI tools (e.g., Otter.ai, Trint, Tactiq) in your dissertation – or that if you have, that you fully declare their use (for a fuller discussion on the use of (generative) AI, see Chapter 2 – *Supervisors, support, and common questions*).

ACKNOWLEDGEMENTS

Your acknowledgements section allows you to recognise and thank those who have supported your dissertation – family, friends, supervisors, gatekeepers, host organisations, participants, etc. While a somewhat peripheral part of your document overall, it is important to think about who to acknowledge and how much detail to add explaining why you are thanking them. Lumping everyone together and saying 'Thank you to everyone who supported me with this dissertation' feels rather half-hearted and superficial, while leaving out someone who has offered significant support can cause unintentional offence.

Some useful prompts to identify those to acknowledge and the brief rationale to include for each might be the following:

- Who has supported you throughout your degree programme and this module (financially and/or emotionally)? – this may be family, friends, staff, etc.
- Who has helped with steering the development and delivery of the project? – commonly your supervisor(s) but there may be others.
- Who has helped in the writing process with proofreading or commenting on sections? – perhaps a family member, a course- or housemate, a friend.
- Who was involved with the research itself? – participants, gatekeepers, partner, or host organisations.

Considering these prompts which of the following three examples feels most suitable:

> I would like to thank my supervisor, Dr Fence, for their guidance throughout the dissertation journey, particularly in analysing the survey data. I am indebted to the residents and wider community members of Little Hebden who participated in this research and provided their time, energies, and insights – I hope my analysis does justice to these. My thanks also to my family, friends, and housemates who have supported me during my dissertation – your provision of tea, chats, and encouragement were invaluable.

> Thanks to my family and housemates for getting me through this – I owe you! Thanks also to my supervisor and other staff in the School.

> Huge thanks to my housemates who have kept me sane throughout my degree, and the dissertation in particular. Mum, dad, and the rest of my family – thank you so much for believing in me and supporting me while I've researched and written this dissertation: shout out also to Humphry the cat and Larry the dog! Many thanks too to my wonderful supervisor, Prof Parsons, who not only helped me develop the initial idea but made sure I kept on track with each stage of the data collection, analysis, and writing. It goes without saying that I owe a huge debt of gratitude to the rangers in the Peak Park who granted access to the landscape and supported with the location of sampling points, maintenance of equipment, recording of results, and the many conversations on music, films, and sport that filled the time walking between sampling sites.

The first is succinct, identifies some specific (groups of) people to thank for identifiable and discrete contributions, and while adding a little personal note retains a professional tone overall. The second is abrupt with no sense that the contributions are genuinely being acknowledged. The third is much longer, veering into the whimsical and irrelevant, and providing an over-abundance of (unnecessary) detail. Typically, the tone and style of the first example would be the most appropriate for a dissertation.

ABSTRACT

Your abstract summarises the key details of your entire dissertation. Your department will have guidance on specific word limit for this section, but it is typically in the region of 250–350 words. Within this, you need to communicate the following:

1 the context and background to your project;
2 what it is you have explored or studied (what question/problem have you tackled);

3 why this question/your dissertation matters (locating your specific study within broader literature and ideas);
4 how you did your research (research design and methods);
5 what your key results, findings, and conclusions are.

These five aspects provide the basic structure for the abstract, although the exact flow of the first part of the abstract will vary depending on personal preference and topic focus. You may find, for instance, that starting with your specific focus and then acknowledging the broader content works well, or it may be more effective to start with the big picture rationale to justify your specific topic. Once you have set the scene, move on to explaining how you did your research before outlining the key findings and conclusions, and the relevance and importance of these. You are not writing a murder mystery novel – you don't want to keep the reader guessing about your main findings, so set these out clearly and succinctly in your abstract. Ultimately, your abstract should 'sell' your dissertation to the reader: giving them a reason to want to invest their time in reading your work while providing a precis of the full document.

You should only write your abstract once you know what your main argument is and findings are. Use the writing of your abstract as an opportunity to check that your dissertation delivers what you think it does and ensure your work consistently builds this core narrative. At this stage it can be helpful to ask a house- or course-mate to read your abstract and then summarise in their own words what they think your dissertation is about, why it is important, how it was done, and the key findings. If their summary is notably different from what you think your dissertation is about then you need to check and edit either the abstract or the main body of your dissertation to more clearly communicate what it is you have done and found out.

WRITING YOUR ABSTRACT

Your abstract may contain some key phrases or sentences from other parts of your dissertation. A useful way of beginning to draft your abstract can be to copy a few individual sentences from the main text to scaffold your abstract. You should then edit these sentences as necessary and add linking content between them to provide a coherent outline to your dissertation. You abstract should not contain quotes, references/citations, acronyms, or technical jargon.

To give you a sense of how to compose a strong abstract, consider the following two example abstracts.

> Microplastic pollution is an environmental issue. Microplastics include plastic debris typically smaller than 5 mm. These pollutants come from a range of sources – from exfoliating facewashes to degrading plastic bottles. Their transport through and accumulation in soil and water can have impacts on plants and animals – including those in the food chain. This means there is a risk to humans from eating plants and animals that are contaminated with microplastics. Looking at three rivers and one coastal bay, this dissertation looks at how much microplastic there is in this environment.

> Microplastic (plastic debris smaller than 5 mm) pollution is an alarming environmental issue worldwide – affecting terrestrial and marine ecosystems. Of particular concern is the impact of microplastic pollution on coastal ecosystems, and the ingestion of these pollutants by marine species. Given the important role played by seafood in the human diet, the consumption of contaminated foods presents a clear danger to human health – particularly among communities whose primary protein source is from seafood. It is thus vital to quantify the levels and sources of microplastic pollution in such environments. This research investigates the levels of microplastic pollution in St Anthony's Bay and its three main rivers, using measurements of microplastics present in both water and sediment. Using visual counting and Infrared Spectroscopy of samples collected from the fieldsite over a 12-month period, the results indicate an accumulation of microplastics in St Anthony's Bay with an apparent seasonality linked to precipitation levels. While the results did not indicate a significant relationship between mean particle size and levels of accumulation, it was notable that the river with the most urban catchment area consistently returned the highest levels of microplastics in both transport and sediment.

The first example provides some detail as to the importance and focus of the work, but in a very disjointed and unconvincing way. The connections between the 'big picture' issue and the specific case study are not clearly explained or justified, and there is no sense as to how the research was done – nor of the key findings. This example does little to convince the reader of the quality of the research being presented. The second provides a more coherent narrative, linking a broader concern to the case study, to the methods and to the key findings, providing the reader with an impression of professionalism and competence and thus of the quality of the research being presented.

LIST OF CONTENTS

Your list of contents provides a simple overview of the structure of your dissertation to aid the reader in finding individual sections of your work *and* seeing the overall structure. The key information here includes the chapter headings, sub-headings, and page numbers.

If you have not already done so, use this as an opportunity to think about the wording of your chapter headings and sub-headings: are they clear and concise? Do they appropriately describe the content of that section? Do they give a clear sense of the specific focus and content of the chapter? Rather than using generic terms such as 'Literature review' or 'Results and discussion' could you effectively use a more descriptive title to outline the substantive focus or argument of the chapter? Some examples of how this might work can be seen in the example contents lists over the next few pages.

Remember to number as well as title each chapter and sub-heading in sequential order, chapter by chapter: within chapter 3 (for instance), you would number sub-headings as 3.1, 3.2, 3.3, etc. You may on occasion find it helpful to use a second layer of sub-headings (3.1.1, 3.1.2, etc.) but be careful not to over-complicate your dissertation by using too many sub-headings as this can lead to a disjointed flow and structure.

Your list of contents should include details of all the content that follows in the rest of the dissertation – including the list(s) of tables and/or figures, list of acronyms, as well as substantive chapters and all appendices. For each entry (unless there is other guidance in your department) include the starting page number. Typically, your front matter (table of contents, etc.) pages are numbered with Roman numerals (i, ii, iii, etc.), with the main body and end matter using Arabic numerals (1, 2, 3, etc.). This should be a reminder to ensure you have added page numbers to your dissertation document. In MS Word, you can use the 'Format Page Numbers' function under 'Insert' → 'Page Number' to modify where your page numbers start from and to add section breaks so your front matter might be numbered using Roman numerals (i, ii, etc.) and your main section using Arabic numerals (1, 2, 3, etc.). It is also helpful to add page breaks between chapters so that each chapter starts on a new page.

SIMPLE CONTENTS LIST

This simple list of contents is not particularly helpful to the reader, beyond being able to see you have a formulaic set of chapters and where to find the first page of each one. It is a missed opportunity to convey to your reader what your dissertation is about and your understanding of how the narrative of your work develops.

DETAILED BUT POORLY FORMATTED CONTENTS LIST

While this list gives a more detailed breakdown of the flow of the dissertation and lists key sub-headings within the work, page numbers are missing in places and the differentiation between levels of (sub)heading are not easily identifiable.

DETAILED AND CLEARLY FORMATTED CONTENTS LIST

This list, while longer, provides further detail on sub-headings – making use of fuller section titles, numbering of sub-headings for ease of navigation (along with page numbers for each (sub)section). Using bold and offset formatting further differentiates chapters from (sub)headings. This gives a much more professional impression and helps the reader identify the flow and focus of the work.

You can compile your list of contents (and list of tables/figures – see next section) manually or using the in-built function of your word processing software. If compiling manually, you will need to copy the chapter numbers, headings, and relevant sub-headings along with the relevant page numbers into a list. Be sure to double check the details are still accurate following any final editing and pagination. Alternatively, you can use the 'Table of Contents' functionality within your word processor (in MS Word this can be found under the 'References' tab) to produce and update your contents list based on an automatic detection of headings. For this to work you need to use the 'Styles' function and ensure your headings and sub-headings are formatted using the appropriate style.

LIST OF TABLES/FIGURES

These lists contain details of all tables and/or figures within your dissertation (including any in appendices). Depending on departmental guidelines, you might have two lists (one for tables, one for figures [i.e., photos, infographics, maps, diagrams, charts]) or one combined list. Number your tables and figures sequentially by chapter – i.e., Tables 3.1, 3.2, and 4.1; Figures 3.1 and 3.2. This helps the reader identify the location of a particular table or figure in your work and reduces the risk of errors in numbering creeping in during the editing process. Ensure each table or figure is referenced in the main text, and that each has a succinct but sufficiently detailed caption. As with the process of compiling a table of contents, you can compile your list(s) of tables and figures manually or using the functionality within your word processing software.

LIST OF ACRONYMS

This list provides an alphabetically organised rundown of all acronyms used in your dissertation and details of the full name of the entity they refer to. The purpose of the list is to provide the reader with an easy means of checking the meaning of an acronym whenever it is used within the dissertation. You must always provide a definition of the acronym the first time you use it in the text as well as in the list of acronyms. The simplest and most reader-friendly presentation of a list of acronyms is a simple two column list:

ANC	African National Congress
ANT	Actor network Theory
BAS	British Antarctic Survey
NERC	Natural Environment Research Council
WTO	World Trade Organisation

CHAPTER SUMMARY

The front matter for your dissertation should be straightforward to piece together. However, as this chapter has outlined, it is important to avoid rushing this material and treating it purely as a formulaic process. These pages are your chance to make a positive first impression on your marker: do not waste it. Spending a bit of time to ensure this content is well presented and thoughtfully developed is time well spent. Similarly, the time spent ensuring your citations and reference list are accurate and correctly formatted is invaluable in ensuring you leave your marker with a strong final impression – as the next chapter outlines.

12

STEP BY STEP

Referencing and reference lists

Throughout your dissertation you will be citing literature to indicate where you have drawn ideas and inspiration from, and to illustrate your understanding of the research field. Ensuring good practice in citations and referencing is crucial to avoid unintentional plagiarism, to substantiate and evidence your argument, to locate your contribution to knowledge, and to ensure your dissertation is presented to a high standard. This *step-by-step* chapter revisits the importance of correct referencing, provides an overview of common referencing styles and approaches, and introduces common reference management software.

THE IMPORTANCE OF REFERENCING

As with all academic work, correct referencing in your dissertation is essential. Your dissertation will have been completed over a relatively prolonged period of time, involving multiple drafts and edits, and with associated risks of errors and oversights in citations creeping in. Throughout this process it is vital you maintain good notetaking practices (clearly differentiating direct quotes, summary notes, and your reflections for each source) and ensure material this then appropriate cited, and citations are not accidently lost or deleted during the (re)writing process. Furthermore, poorly presented citations and reference lists can be viewed as a sign of a rushed piece of work, casting doubt on the quality of your dissertation and lost marks as marking criteria often make explicit reference to the quality, accuracy, and presentational standard of citations and reference lists.

DOI: 10.4324/9781003584315-12

Returning to the guidance in Chapter 2 (*Supervisors, support, and common questions*) relating to the use of Generative AI – never use materials verbatim from GenAI platforms in your dissertation. Linked to this, you must always return to and engage with the sources cited in any generated content as you need to read and interpret these yourself. This also ensure you avoid falling foul of invented or 'hallucinated' references that are a common problem with AI-generated material and can lead to significant academic misconduct penalties if discovered.

IN-TEXT AND FOOTNOTE CITATION APPROACHES

While multiple specific referencing styles exist, these can be grouped into two main approaches: in-text and footnote. Both approaches are used in the social and physical sciences, and many departments have their own guidelines as to which you should use. While some readers find in-text citations hinder the flow of the text and dislike

ILLUSTRATIVE COMPARISON OF IN-TEXT AND FOOT-NOTE CITATION APPROACHES

Postage stamps are one such quotidian object, giving material and symbolic expression to government's national imaginings for domestic and international audiences (Raento, 2009; Wallach, 2011).

Above, an example sentence with Harvard-style in-text citation – with full reference details then added in a final reference list, compared with (below) footnote style referencing for the same sentence.

Postage stamps are one such quotidian object, giving material and symbolic expression to government's national imaginings for domestic and international audiences.[1]

Note

1 Raento, P. (2009). Tourism, nation and the postage stamp: Examples from Finland. *Annals of Tourism Research*, 36/1; Wallach, Y. (2011). Creating a country through currency and stamps: State symbols and nation-building in British-ruled Palestine. *Nations and Nationalism*, 17(1).

the lack of full reference details at the first point of citation, others prefer this limited inclusion of information within the text and consolidated reference list at the end of a document. Whichever approach you adopt, ensure it meets your department's requirements and be consistent throughout your dissertation.

IN-TEXT CITATIONS AND REFERENCING

In-text citations – of which the Harvard style is a common example – contains key details of the author, year of publication, and – where a direct quote is used – the page number of the quoted material within the line of text. While this information is typically all included within parentheses, there are times when the authors name is included as part of the sentence and not in parentheses – for example:

> Recent writing about the geographies of populism and immigration has addressed the influence of right-wing populism in the United States in comparison with the UK's vote for Brexit (Nagel, 2009).
>
> In her recent work on populism in American politics, Nagel (2009) has critically reflected on the drivers behind the rise in anti-immigrant sentiment.
>
> In writing about right-wing populism in the United States, a key theme has been how Donald Trump's 2016 election victory indicated "a breakdown of the liberal consensus that has guided immigration policy since the 1960s" (Nagel, 2009: 12).

Where there are two authors on a paper, both names would be included in the in-text citation – for instance (Ehrkamp and Nagel, 2017). Where there are three or more authors of a paper, the in-text citation is the lead author's surname followed by 'et al.' For example, if we were discussing research on early-Holocene warming and glacial ice melt we might cite (Langdon et al., 2010), with the full details included in the reference list as follows:

> Langdon, P. G., Leng, M. J., Holmes, N., & Caseldine, C. J. (2010). Lacustrine evidence of early-Holocene environmental change in northern Iceland: a multiproxy palaeoecology and stable isotope study. *The Holocene*, 20(2), 205–214.

A document using in-text citations requires a consolidated reference list of references at the end of the main text – *not* a bibliography. A bibliography is a list of possible readings on a specific topic or of all the sources you have read. A reference list only includes those sources you have read *and* cited in your document. Unless otherwise instructed by your department, include a reference list not a bibliography.

REFERENCES STYLE GUIDES

Multiple style guides exist and provide templates for the formatting of citations and reference lists, including MLA, APA, Chicago, Harvard, and Vancouver. Your department will likely have a set style guide for you to use – please adhere to this. If they don't then adopt one of the common ones and use this throughout. Regardless of which style is used, they all contain the same core information to allow a reader to identify and locate the sources you have utilised. While some templates have removed the need for the 'place of publication' information for books, this is typically still requested. Where possible you should also include DOIs (digital origin identifiers) or URLs for accessing online resources (including academic journal articles).

While the exact formatting for your reference list will vary depending upon your department's guidelines, the examples below illustrate the information commonly required:

Journal article:

Fox, A., Ziervogel, G., & Scheba, S. (2021). 'Strengthening community-based adaptation for urban transformation: managing flood risk in informal settlements in Cape Town'. *Local Environment*, 28(7): 837–851. https://doi.org/10.1080/13549839.2021.1923000

Book Chapter:

Magaya, S., & Fitchett, J. M. (2022). 'Approaching positionality in research on indigenous knowledge systems', in Ebhuoma, E.E., & Leonard, L. (eds) *Indigenous Knowledge and Climate Governance: A Sub-Saharan African Perspective*. Cham: Springer International Publishing, pp. 81–93.

Book:

Goodfellow, T. (2022). *Politics and the urban frontier: Transformation and divergence in late urbanizing East Africa*. Oxford: Oxford University Press.

Thesis:

Leslie, R. (2024). *Quantification of solubility trapping in natural CO_2 reservoirs using gas geochemistry and reservoir modelling*. PhD Thesis, School of Geoscience, University of Edinburgh.

Webpages:

RGS-IBG. (2024). *Responsible and ethical fieldwork: working in/on an environment*. https://www.rgs.org/research/higher-education-resources/ethical-fieldwork-working-on-environment (Accessed 11th December 2024).

Newspapers:

Myers, P., and Smith, D. 2010. 'As football begins, West is warned: don't stereotype whole of Africa'. *The Guardian*, 11 January 2010.

Working Papers/Reports:

Dookie, D. (2024). *Essential role of climate information systems for early action interventions and resilience-focused decision-making*. ODI/RESI Working Paper. Available at https://odi.org/en/publications/role-of-climate-information-systems/

FOOTNOTE CITATIONS AND REFERENCING

Footnote-based referencing approaches use an in-text notation – typically superscript numbers in sequential order throughout a document – linked to reference details included in a footnote at the bottom of the page. The first use of a source would have full reference details included in the footnote, with duplicate entries then using abbreviated reference details (typically the first author's surname and abbreviated title). For example, if the first footnote reference were as follows:

Abrahamsen, R. and Williams, M.C. 'Securing the city: private security companies and non-state authority in global governance', *International Relations*, 21 (2007), pp. 237–253.

Then, a subsequent entry would be the following:

Abrahamsen and Williams, 'Securing the city'

Footnote referencing approaches typically do not include a consolidated reference list at the end of the document. Whichever approach you use, you may find it useful to use reference management software and 'cite-while-you-write' plug-ins to organise and maintain your references and ensure accuracy in your citation practices.

REFERENCE MANAGEMENT SOFTWARE

Managing and updating reference lists and citations in a longer document such as a dissertation can be a time-consuming and frustrating task. Reference management software collates details of your sources in a library and allows you to manage your reference list and citations within the document through a 'cite-while-you-write' add-on.

These programmes require time to learn how to use and identify or create the correct 'output style'. You also need to build your reference library: you can typically import these directly or add details manually. Once sources are in your library you can typically add your own notes as well as pdfs of the sources. Once you have begun to build a library you can activate a cite-while-you-write plug-in allowing you to add citations into your document and automatically build your reference list (using the chosen output style which should ensure correct and consistent formatting).

Four of the most commonly used by academics and students are EndNote, Endnote Web, Zotero, and Mendeley (Table 12.1). Before using any of the software, check which one(s) your university has a licence for (if required) and what support materials are available (e.g., your university library will likely have some

Table 12.1 Summary of Common Reference Management Software

	Endnote	*Endnote Web*	*Mendeley*	*Zotero*
Can assess and synch your reference library across multiple devices		X	X	X
Can work offline and store reference library on your computer	X		X	X
Can attach files	X	X	X	X
Allows you to search full text of attached PDFs	X		X	X
Allows you to edit output styles	X			X (but complicated)
Cost	Paid for (but check if your university/ library has a licence you can use)	Free	Free	Free
Works with MS Word and/ or Google Docs	MS Word; Google Docs	MS Word	MS Word	MS Word; Google Docs
Strengths	Very effective for large reference libraries; provides a vast array of output styles which are also easily edited to fit department requirements.	Easy to use across multiple devices; outputs styles are editable to fit department requirements.	Automatically syncs across devices and draws citation data from PDFs	Easy to use; collects and organises sources rapidly; user friendly and allows tags, annotations, and notes

(*Continued*)

Table 12.1 (Continued)

	Endnote	*Endnote Web*	*Mendeley*	*Zotero*
Weaknesses	Does not sync across multiple devices	Requires internet access to use (cannot work offline); limited number of output styles provided; limited storage space	Cannot edit output styles to match departmental requirements; limited storage space	Tricky to edit output styles if needed to match departmental requirements; limited storage space

accessible resources for some platforms). If you opt to use one of these programmes, it is advisable to

1 use their inbuilt systems to automatically capture the reference details (author, title, etc.) where possible to minimise the risk of typographical errors;
2 check the available output styles before starting to ensure either a match to your department's style guide requirements or that you can quickly edit one to meet these requirements;
3 be consistent from the outset in adding all source materials to the library.

(HOW) DO I CITE...

You should be well versed in good citation practices and requirements from previous coursework assignments. However, your dissertation may require you to engage with and cite different types of material to those used previously. Some of the more typical considerations are addressed here.

INTERVIEW TRANSCRIPTS AND/OR PERSONAL CORRESPONDENCE

Unless you are using materials that are already held in an accessible archive or library (e.g., British Library), these should not be included in the main reference list. The reference list provides details to allow the reader to find the materials being cited so does not include details on private materials such as non-archived interview transcripts, personal correspondence, or your research diary.

An entry for archival materials would include four main elements – the title of the file (or a short description if untitled – e.g., Letter from Albert Mattock to Esther Jackson), the name of the collection in which the material is held (e.g., National Fairground Archives), the reference code or identifier, and name and location of the repository (e.g., British Library, London, UK).

DATASETS

If you have created your own dataset then you do not include this in your reference list. If you are using an existing dataset then you do include these details in your reference list – including the digital object identifier if this exists and details of the author, data, title of data(set), the date accessed and where the data(set) is available from. As an example:

The World Bank. (2024). *World Development Indicators*. Washington, D.C.: The World Bank. Accessed 17th November 2024, available from https://datacatalog.worldbank.org/search/dataset/0037712

CITING A CITATION

Sometimes you come across an idea or quote that is cited in a source you are reading but is originally from another document as in the following excerpt from Philip Steinberg and Kimberley Peters' paper 'Wet ontologies, fluid spaces' in *Environment and Planning D*.

> This viewpoint is mirrored in the anthropological writings of Claude Lévi-Strauss, for whom the ocean is "a diluted landscape" with an "oppressive monotony and a flatness" that fails to hold qualities to enliven the imagination (1973, pages 338–339), and Roland Barthes's depiction of the sea as a "non-signifying field [that] bears no message" (1972, page 112).
>
> (Steinberg and Peters, 2015: 249)

In this situation, if you wanted to cite the quote from either Lévi-Strauss or Barthes what should you do? Whenever possible you should return to the original source material for the citation and read this, and then cite this original source material. In this instance, that would mean finding and reading translated (unless you are fluent in French) copies of Barthes's *Mythologies* and Lévi-Strauss's *Tristes Tropiques* and developing your own reading of these quotes and ideas.

Sometimes this is not feasible – for instance if the original source is in a different language, or is from a book or other source or archive that you cannot access. In this situation, you need to clearly identify in your writing that you are citing someone else's citation. For example:

> While the ocean can be described as having an "oppressive monotony and a flatness" (Lévi-Strauss, 1973: 339, cited in Steinberg and Peters, 2015: 249) this assumes a particular perspective.

The crucial aspect is that while the original source of the quote is acknowledged (Lévi-Strauss), the citation does not claim to have read this but identifies that it is drawn from (and reliant upon) Steinberg and Peters' reading of Lévi-Strauss. In your reference list you then include details of the source you read and cited – not the originally cited source. Using the above example again, you would include the Steinberg and Peters paper in the reference list, but not Lévi-Strauss's book.

This may seem like a minor and pedantic concern, but it is important for several reasons. The first is academic honesty and integrity – it is malpractice to cite works you haven't read and can result in allegations of misconduct. Secondly, you are also reliant upon the interpretation of the original source material by the author you are reading: if they have misread or misinterpreted (or selectively interpreted or quoted) the original source material, you will reiterate this error and risk being penalised for it.

PHOTOS AND OTHER IMAGES/GRAPHICS

If you have used a photo, map, table, graph, or other image that you have taken from another source, you need to include the details of where this material has been taken from and provide the relevant attribution/citation at the end of the caption for the figure or table. Additionally, be conscious of possible copyright restrictions: while many sources now have a Creative Commons licence that allows various types of reuse, others are subject to much more stringent restrictions. If in doubt, the best source for information and guidance will typically be your university library and librarians.

If you have designed, developed, and drawn the infographic, map, graph, illustration, or have taken the photo yourself, then it is useful to state this. There are main two reasons for this – the first is so you get credited for the work you've put in, and secondly to stop the examiner asking 'where has this come from, you need to provide the source'. The simplest way to provide this information is to "Source: Author's own photo" or "Source: Author's personal sketch" or equivalent wording in the relevant caption.

DOES THE REFERENCE LIST COUNT TOWARDS THE LENGTH LIMIT?

You must check your departmental guidelines on this as instructions vary between institutions. Often the reference list is not included in the length limit as this would act as a disincentive to wider reading and full referencing in the document, but this is not always the case.

CHAPTER SUMMARY

Ensuring consistency and completeness in your citations and references is vital not only to ensure you demonstrate good academic practice (and avoid accusations of or penalties for malpractice) but also to meet related requirements in marking criteria. Ensuring this content is correctly (and consistently) formatted and presented also reassures your marker that you have taken your dissertation seriously and are presenting a competent and reliable piece of research.

13

STEP BY STEP
Appendices

An appendix (singular – appendices plural) comes at the end of your dissertation and contains additional information or data that is relevant but not essential to the main text. Your dissertation should be understandable, logical, and make a complete and coherent argument without the materials in the appendices. In some cases you may be required to include particular information (ethics approvals, interview schedules or transcripts, scientific instruments or coding, etc.) and in others you might have materials that the reader may want to engage with but which would be a distraction in the main text.

The type of information included, and importance of providing this, will vary from dissertation to dissertation, and between sub-disciplinary areas and methodological approaches. While appendices are typically excluded from word limits they are *not* an opportunity for you to use to circumvent length limits. If your marker thinks you are trying to use appendices to do this, the content will either be discounted or subject to overlength penalties. The rest of this *step-by-step* chapter sets out guidance on when and how to use appendices, addressing commonly asked questions from the norms of naming appendices to what kinds of materials to include.

WHEN TO USE APPENDICES

One frequently asked questions of supervisors is 'should I have appendices?' or 'should I put X in my appendices?' Many dissertations benefit from having appendices, but this is not universal. You need to think about why you need them and what useful

DOI: 10.4324/9781003584315-13

information will be contained therein. You should only include any appendices if they serve a useful purpose. Such content can be understood as follows:

1 Content (typically data) that be too long and detailed to fit within the main body of the dissertation without significantly disrupting the flow and readability.
2 Useful additional resources or information that would be a distraction in the main text.
3 Meeting specific departmental requirements, such as your ethics application or approval, list of interview questions, and anonymised interview transcripts (if relevant).

Each appendix should be self-contained, covering an individual topic, issue, or set of data, and connect to (i.e., be referred to in) the main body of your dissertation. You should not use appendices to simply demonstrate how much data has been collected – there needs to be a clear and logical rationale for the materials to be included. An obvious sense check on this is to think about where in the main body of your dissertation you direct the reader to look at each appendix for further relevant information. If you don't reference them then (i) the reader won't know where to look for the information they contain and (ii) the content is probably not needed and can be deleted.

On the flip side, if the content of an appendix is integral to the main narrative of the work it should be in the main body of work. A common error by students – particularly those working with visual data such as photographs or maps – is to place these images into appendices and then repeatedly reference or analyse these materials without including the image alongside the discussion. Any image, map, photo, graph, or specific data table directly mentioned or referenced in the main body of your dissertation should be included in the main body (unless this would significantly disrupt the readability of the text in which case locate this in appendix).

The following scenarios illustrate these considerations.

You have an extensive series of data tables that illustrate the changing vote share per constituency over a series of general elections which have been used to generate a small number of maps or infographics. Where do you place the maps/infographics and the data tables?

The maps or infographics will likely be key to your discussion so would be in the main text and directly referenced and analysed. Any additional maps or infographics that are not directly referenced and discussed but which provide broader information and detail may be added as appendices. The data tables are useful additional materials that have a purpose (to allow a reader to explore further if desired and see that your data is robust and reliable) but take up dozens of pages, so these would be located in an appendix. You may choose to include one or two example data tables in the main text to illustrate your data collection and analysis processes or if you draw directly from statistics and data within these tables.

You have a series of 45 photos from a transect walk showing changing landscapes. You discuss 12 of these in detail in your analysis. Where do you place the photos?

The photos that are discussed go in the main text alongside the relevant discussion. You may want to include another example photo or two in your methods section to illustrate your method and approach. The rest of the images would be more suited to inclusion in an appendix: they provide evidence of the robustness of your data collection and may provide supplementary data for interested readers, but the images are not core to your analysis.

Your dissertation is based on interviews with farmers on their views on rewilding and their practices in relation to wildlife conservation. You have a copy of your ethics approval, your interview schedule, and multiple interview transcripts.

Check your departmental guidance to see if you are required to include any or all these documents. If there is no requirement to include, then ask yourself what is the value added is for the reader? Details of ethics approvals should be held centrally via the approvals process, so is there a need to include here? A copy of your interview schedule can be useful in showing the marker the questions asked and potentially alleviating any concerns regarding leading questions and potential bias, but is often not essential. Finally, unless your department requires you to include transcripts, these are best left out as they are time consuming to edit to ensure anonymity. You might consider including a list of interviewees (anonymised as appropriate in line with your ethics approval and agreements with interviewees) if you have not included this in the main text. If your department does require transcripts are included, then you must comply but check expectations for what to include and ensure you comply with ethics requirements regarding anonymising each transcript prior to submission.

Similarly, unless your department specifies that you attach approved risk assessment paperwork, participant information sheets or consent forms then do not include these. If your department requires you to include these documents be sure that they are suitably redacted/anonymised – including the removal of personal information (names, phone numbers, home addresses, etc.) that should not be shared publicly.

LABELLING AND FORMATTING APPENDICES

As with chapters, use a clear numbering sequence alongside short descriptive titles – as you would do with all chapters in your dissertation. You should include these details at the start of the respective appendix as well as in your list of contents (see Chapter 11 – *Step by step: Front matter*). A common approach would be the following:

Appendix 1: Additional photos from Peak District transect walk (Edale to Hathersage)
Appendix 2: List of and links to audio files of ambient noise for transect walk soundscape
Appendix 3: Background data tables on changing voting intentions, Peak District constituencies (1983–2024)
Appendix 4: Full list of species recorded during transect walk with their Latin and common names.

You should use the same formatting guidelines as the main body of text for your dissertation, and follow the same style for formatting (sub)headings, table and figure captions, and numbering. For instance, if you numbered tables in your main text on a per chapter basis, then tables in chapter 4 of your dissertation would be numbered Table 4.1, Table 4.2, etc. – you might then number tables in your first appendix as Table A1.1, Table A1.2, and then Table A2.1, etc., for appendix 2, and so on.

CHAPTER SUMMARY

Appendices should not be an afterthought to your dissertation. You need to think carefully about whether you need to include these and, if so, the purpose they serve. Be sure to meet any departmental requirements relating to compulsory content (if relevant) and then be strategic with any further content you add. This should always be useful and relevant in supporting the main narrative, but not essential to it. Simply adding tangential content to demonstrate the volume of work undertaken or placing key materials in an appendix that the reader has to jump back-and-forth to will not make a good impression – and you want to keep your audience 'onside', as the next chapter outlines.

14

REMEMBER YOUR AUDIENCE

Keeping your marker onside

As with all writing, it is vital to know who your intended audience is and to write for them. In answering the question of who will read your dissertation, we must then think about what this means for your writing and what formal requirements and unspoken expectations need to be met? While the primary audience for your dissertation is academic, there may be other audiences to consider. This chapter focusses on meeting the expectations of your academic audience before considering who else might be interested in your work – a theme that is continued and developed further in Chapter 15 – *What next? Employability and your dissertation*. First, though, it is important to consider who will read your dissertation.

WHO WILL READ YOUR DISSERTATION?

The harsh truth is that the audience for your dissertation in all probability amounts to you, family members (perhaps), your supervisor(s), and your examiners. If you have worked with a specific external partner, organisation, or community it is possible they will demand or want to see the final output (or, more likely, an abbreviated version of your dissertation).

On an instrumental and strategic level you need to write your dissertation for the one audience that matters the most: the academics marking your dissertation. It is their view on the quality, reliability, and robustness of your work, and the extent to which it meets the marking criteria, that determines the grade awarded. With geography dissertation modules often worth double the credits

DOI: 10.4324/9781003584315-14

of a standard taught module (or even double-double weighted in institutions where Level 3 marks are double weighted compared to Level 2 marks), this mark can have a significant bearing on your final degree outcome.

Double-marking of dissertations is common practice. In some institutions your supervisor will be one of the markers, but not in others. This means that at least one, if not both, of the academics assessing your work will not have been involved in extensive discussion with you about your project and may not have expertise on the specific focus of your work. In short, you cannot assume that your reader has prior understanding or knowledge of your topic, methods, coding, and so on. A common trap that many dissertations fall into is assuming the marker already knows the rationale for key decisions or other details that were discussed with a supervisor. Do not make this assumption and ensure that your dissertation is clear in structure, content, and justification for key decisions such that someone who has no prior knowledge of your work can understand it. In addition, make use of strategic signposting and wording to clearly evidence how your work meets the relevant marking criteria.

ENGAGING WITH AND UNDERSTANDING THE MARKING CRITERIA

Your examiners are looking for a dissertation that is focussed, understandable, well designed, and based on reliable data and robust analysis. They want to see that you have genuinely *thought* about your project from start to finish, and that your argument clearly demonstrates critical thinking and an understanding of the research process. When reading your work, your marker will have various general questions in mind along the lines of:

- Have you addressed the title of your work?
- Have you identified and delivered on well-articulated and deliverable research questions that are clearly justified and located in relation to existing work?
- Do you have command of the conceptual and theoretical literature you are using to both frame and then analyse your project and data?

- Have you chosen the most appropriate methods to collect reliable, relevant data?
- Has this been subjected to robust and appropriate analysis to produce a clear and considered discussion?
- Have you answered the questions, aims, or hypothesis you set out to explore?

Your examiners will then assess your work in relation to the expectations set out in the marking criteria. Assuming your department makes these marking criteria available to you, take the time to read and understand these (including making use of any accompanying glossary that explains the definitions and meanings of some key terms and phrases used in the marking criteria). Ultimately, these are the requirements against which your works is assessed so it is worth understanding them.

Various common expectations and terminology permeate across individual marking schemes (see also Nicholson et al., 2010) and can be grouped into three main arenas: scholarship, independent learning, and communication. Within these themes various terms and foci are apparent, albeit with some overlap between themes:

Scholarship and independent learning:

- evidence of originality and independent thinking;
- well-defined problem/aims/questions;
- awareness and synthesis of existing literature;
- valid research design;
- robust methods and data collection;
- quality of data and analysis;
- ethical conduct of the research and reflections on this;
- justified conclusions.

Communication:

- clarity and precision in presentation;
- clear and substantively error-free writing;
- citations and references are consistently and correctly presented.

The presentation of marking criteria varies between institutions. In the example contained in Table 14.1, the criteria for each grade band (or sub-band in the case of First Class and Fail bands) are

Table 14.1 Example Dissertation Marking Criteria (Excerpt from Full Marking Spectrum), University of Sheffield's Department of Geography 2023/2024

Upper Second Class	60–69	Execution and Presentation: Good execution of a worthwhile idea but with identifiable unfulfilled potential. A sound approach and an effective structure and style. Well illustrated and referenced with correct and consistent reference to illustrations and literature throughout. Very capably written throughout with few errors. Appropriate use of specialist or technical terminology. Expression is fluent and coherent. Research Design: The aim addresses a worthwhile opportunity that is clearly justified and explained. The context and rationale draw upon an appropriate literature that is also used to inform and justify the overall approach and choice of methods. Methodology and Methods: The approach is justified and articulated clearly and a sensible framework is described that specifies how the intended outcomes will be achieved. Specific methods to be used are technically correct, are appropriate, and are satisfactorily justified and described. Results and Analyses: Adequate data of good quality has been collected to address the intended outcomes. The results are presented logically and sound and appropriate analyses are used to test and evaluate the strengths and limitations of the results in relation to the intended outcomes. Discussion and Conclusions: A sound evaluation of the findings and their significance in relation to the project aim and appropriate literature. Any limitations affecting the findings are reasonably well countered and explained. Where appropriate to the dissertation, there is good separation of discussion from the results. Conclusions are fully supported by the results and analyses that have been presented.

(*Continued*)

Table 14.1 (Continued)

Lower Second Class	50–59	Execution and Presentation: Fair execution of a worthwhile idea but with clearly identifiable unfulfilled potential. A reasonable approach and competent structure and style. May lack sufficient organisation and illustration, and there may be errors relating to spelling, punctuation, or referencing. Competently written throughout but sometimes lacking clarity and precision. There may also be errors of fact, terminology, or interpretation. Research Design: The aim addresses a worthwhile opportunity that is reasonably well described but may lack justification. The context and rationale cite key literature but the selection and synthesis of material and the quality of arguments provides a weak grounding and justification. In addition, there is limited application of literature to inform and justify the overall approach and choice of methods. Methodology and Methods: The approach and specific methods are appropriate but are unambitious and may lack justification. The approach is reasonably clear and a reasonable framework is described. Specific methods are mostly correct but occasionally imprecisely described. Results and Analyses: Reasonable data of adequate quality has been collected to address the intended outcomes. The results are presented logically and a reasonable attempt is made to test and evaluate the strengths and limitations using mostly appropriate analyses that are mostly correct. Discussion and Conclusions: A reasonable evaluation of the findings and their significance in relation to the project aim and appropriate literature. Where appropriate to the dissertation, results and discussion may not be clearly separated. Limitations that affect the findings are acknowledged but weakly countered and explained. The conclusion is reasonably well supported by the work presented. *Work that shows upper second-class qualities may fall into this category if it is too short, unfinished, badly organised, or poorly written and presented.*

divided into five themes: Evaluation and Presentation, Research Design, Methodology and Methods, Results and Analyses, and Discussion and Conclusions. Here, the markers are not given a prescriptive weighting of marks between these themes – although this can be the case elsewhere. Nonetheless, the descriptor for each theme outlines a set of expectations to be met for the dissertation to be awarded a mark within the relevant grade band.

Higher grade bands expect greater demonstration of quality in the design, execution, and communication of your dissertation. They will also contain increasingly stringent requirements relating to the communication of your critical comprehension of existing literature, justification of your research topic and research design, the execution of this with appropriate (potentially innovative) methods resulting in good data which is then subject to rigorous analysis leading to fully supported conclusions.

Being aware of the specific language and requirements of your department's criteria allows you to more clearly signal how you meet these expectations. One helpful technique can be to strategically use certain key terminology to highlight what you see as being the originality of your dissertation. Language such as "This work provides novel insights into" or "This paper addresses the gap in existing debates on X through a detailed examination of Y" can be really effective in mobilising claims to originality and gaps in understanding. It is also useful to clearly structure and signpost your dissertation – both in setting out an overview of the work in the introduction, and in using the opening and closing paragraphs of each chapter to (i) highlight the connections and progression between chapters and (ii) clearly communicate the main message or 'take away' from the chapter. These techniques (along with ensuring you have addressed the questions in the following text box) can help your examiner identify how and where you are meeting the marking criteria – and keep them on 'your side'.

KEY SELF-REFLECTION QUESTIONS

- Have you clearly articulated your understanding of the key parts of the relevant framing literature?
- Have you explained how and why your work builds on this?

- Have you set out a clear and achievable research aim(s) and question(s)?
- Have you justified you selection of methods and data collection approach and subsequent approach to data analysis?
- Are these choices appropriate?
- Have you demonstrated that you executed these methods correctly?
- Do you clearly use and present empirical materials or other data?
- Do you explain these data and justify the meaning you derive these materials?
- Are you interpreting your data appropriately and making relevant links to existing literature and previous findings or studies?
- Are you crediting the sources you are drawing upon properly?
- Is your writing clear and precise, and checked for errors in spelling, grammar, and syntax?

KEEPING YOUR MARKER ON-SIDE

Your dissertation marker is going to be juggling other demands on their time, so it is in your interest to gain their attention and buy-in to your work while also ensuring your dissertation is straightforward to read and grade. This will generate goodwill that can filter into the marking process. In other words, you want to keep your marker 'on-side'. How might you do this?

The answer is not bribery and corruption, nor gifts of coffee and chocolate (that way lies disciplinary action). What you can do is to make life easy for your marker. Despite what you may believe, academics are not seeking a reason to fail your work or mark you down. More often we are looking for reasons to award you more marks – but we can only do this in accordance with the marking criteria. So, the first thing to do is to engage with the marking criteria and ensure your work addresses and signals how it meets the requirements that are set out, from presentation to conceptual framing, from methods through analysis to conclusions.

This is where asking a course-mate to proofread your dissertation can also be helpful (and to offer to do the same for them, it is not a competition). As well as asking them to check for any obvious typographical, spelling, or grammatical errors, you can also ask them to identify if any sentences don't make sense or are incomplete, if it's unclear why a section or paragraph is included, and whether it is clear how you meet the marking criteria. Aim to do this a week or two before the submission deadline – and you have ample time edit and improve your work as necessary.

Returning to the ideas covered in Chapter 4 (*Writing well*), you can also help to keep your marker onside through some of the practices Gary Provost (2019) identifies to ensure people like your writing. Specifically, be sure to

1 name your sources;
2 provide useful information (background contextual detail, etc.);
3 use quotations and/or other data to illustrate and support your argument;
4 show your opinion (this is your dissertation so convey your argument/opinion throughout);
5 use anecdotes and examples where relevant and appropriate to illustrate your argument;
6 frame the work with a clear, strong title (and, to paraphrase the famous Ronseal slogan of 'Does exactly what it says on the tin' – make sure your dissertation does exactly what it says in the title).

On the flip side, try to avoid the use of too much technical jargon and cliches. A key failing of many academic papers (and some dissertations) is that the author tries too hard to sound intelligent and fails to clearly communicate their argument. Overly complex writing can be read as a sign that you are not in command of the literature and thus struggle to make *your argument*. It is vital you mobilise literature to support but not dominate or lead your argument and analysis – you want the examiner to quickly identify that you know how your work builds on existing thinking.

ORIGINALITY

In the context of your dissertation the need to demonstrate originality boils down to 'what is new in your dissertation?' Remember, this is an undergraduate dissertation so the degree of 'newness' needs to be kept realistic: you are not expected to produce a treatise proposing an entirely new theoretical framework, nor to identify a hitherto unidentified sub-glacial process that determines meltwater velocity. As outlined in Chapter 1 (*Thinking about your dissertation*), your claim to originality can be much more modest, and typically fits into three main categories:

1 Using a new method or new data to explore a question – this may be in the form of a new case study on a particular topic or issue, adopting a different method to develop further understanding and insight.
2 Utilising a new perspective on a topic or testing existing ideas in a new way.
3 Incrementally building on previous work on a topic – the classic 'standing on the shoulders of giants' approach to develop further empirical or conceptual insights that try to advance an existing debate.

With 'originality' being a key assessment criteria it is really important to demonstrate to your examiner how you meet this need, and to justify how you deliver on this.

FIRST IMPRESSIONS COUNT

We develop our first impression of somebody in less than a second (Willis and Todorov, 2006). While a first impression from a dissertation takes a little longer to form, this still matters. As previous chapters have outlined if you present a marker with a poorly formatted and badly written dissertation and you have more of an uphill struggle to get them to follow your key argument and analysis than with a clearly formatted and written piece of work. In other words, help us (the markers) to help you (get a better grade).

What seem to be the minor details really matter – if your department provides a template for the cover sheet then be sure to use this. Follow the instructions you are given in relation to what information to include where, not least how to name you file for submission. Adhere to your department's style guide in terms of font type and size, line spacing, and referencing. A well-presented and easy-to-access document gives a positive impression to your marker while demonstrating that you meet relevant marking criteria relating to presentation and style.

Remember the guidance earlier in this book about writing well and telling a 'simple story elegantly'. Clear, active writing and a robust, consistent red-thread will help keep your marker on-side as this will make it easier for them to readily identify your synthesis of literature, originality of argument, depth of analysis, and so on. This then ties into ensuring your dissertation is well organised – can the examiner quickly see the flow and focus of the work from a detailed and comprehensive (and well formatted) contents page? Are the (sub)headings clear and logical? Is it clear where one chapter ends and the next starts? All of these little things matter.

FINAL IMPRESSIONS ALSO MATTER

Just as the opening page(s) and first glance at the main body of your dissertation matter, so too does the final impression of your work. The last section of the dissertation that a marker looks at will have a final framing effect as they draft your feedback and mark. This not only means that you conclusion needs to be succinct and sharp, but your reference list needs to be accurate and correctly formatted (see Chapter 12 – *Step by step: Referencing and reference lists*). Think back to the marking criteria – it's likely that there will be a component of those criteria that talks about the presentation of your citations and reference list: these are easy marks to pick-up (and to lose). So, factor in the time to ensure your reference list is accurate (all cited sources in the text are included and nothing else) and presented correctly (references are in the required order, and the necessary information for each reference is present and formatted according to your department's style guide).

CROSS-REFERENCING AND AVOIDING DUPLICATION

Strategic use of signposting and cross-referencing within your dissertation can minimise duplication and repetition. The marker does not need to read the same content in the introduction, literature review, and discussion sections. You may need to provide very succinct summaries of reminders in places, but at other moments you might be able to simply signpost connections between sections.

For instance, if you include the research questions in your introduction, rather than repeating them later you can say "In answering my research questions (see Introduction)".

If you have used appendices to include additional materials, cross-referencing these in the main text is key. If you don't, why would your marker spend the time to look at them? Equally (and as covered in Chapter 13 – *Step by step: Appendices*), if you are constantly referencing appendices in the main text, your marker will get frustrated having to constantly shift back-and-forth between sections of the dissertation. If you repeatedly reference an appendix, then reconsider whether that material should be included in the main part of the dissertation. This can often be the case when images, maps, or graphs are included as appendices when they are key to the discussion being presented.

NOT JUST YOUR MARKER: THINKING ABOUT OTHER AUDIENCES

Your examiners are the primary audience for your dissertation. But they need not be the only audience – for many dissertations there are other potential audiences you might want to engage. This is particularly the case for projects that have been based on or with an external organisation – a business, a civil society or community group, a school, etc. – where there may be an expectation or moral imperative to 'give back' and to share findings with them. These audiences are not going to want to – nor have the time to – read a full academic dissertation.

In such situations you need to think about how you might translate your research findings into a shorter, more accessible output. This may be as a policy brief, a research note, a one-page summary, or a poster. The decision on how to present your findings for a partner organisation should be made in discussion with the partner: what will be most useful/relevant for them?

Another audience for your dissertation is potential employers – while the topic of your dissertation may be tangential to or disconnected from the interests of a potential employer, your dissertation can be a key resource nonetheless in your application. Again, an employer won't read a dissertation but being able to communicate the key findings and message in a concise manner allows you to demonstrate multiple relevant transferable skills. Depending on potential employment field/employer and role applied for, think about how you might distil your dissertation and communicate in an appropriate manner – can you distil to a two-page policy brief? A one-page press release? A (short series of) social media post(s)?

In each of these situations you will need to think about how to condense your dissertation into a much shorter summary and consider whether – depending on the audience – you want to focus in on a particular part of your work and findings or to summarise your whole argument. Some initial guidance and advice on this process can be found in the next chapter, *Next steps: Employability and your dissertation*.

CHAPTER SUMMARY

There may be several audiences for your dissertation, the most important one on an instrumental level is your examiner(s). It is in your interests, therefore, to keep them onside and make it as easy as possible for them to give you credit – marks – for the work you have done. In part, this is realised through clear and accessible writing and clarity of argument. It is also achieved through paying attention to and clearly demonstrating how you meet the marking criteria used in assessing your dissertation. Once you have ensured you have done what you can to keep your marker onside you might then begin to consider how else you might utilise your dissertation findings for other audiences – a process considered in the next chapter.

15

WHAT NEXT?

Employability and your dissertation

Your dissertation is the keystone of your degree and a crucial resource for you in the job market, a document (and process) that demonstrates an array of research, technical, communication, and transferable skills. This chapter outlines various strategies for translating your dissertation into examples or illustrations of skills and competencies in job applications and interviews, and how to confidently and competently translate the skills and knowledge developed during your dissertation into the language of employability.

HOW TO TALK ABOUT YOUR DISSERTATION

While your dissertation can be a key resource in demonstrating how you meet job application criteria and in responding to interview questions, you need to be strategic. In other words, don't use your dissertation as evidence to respond to every question – this will get repetitive. Rather, be selective in using your dissertation to evidence specific skills, attributes, and knowledge relevant to the job you are applying for. To do this effectively you will need to map your dissertation experience to different skills or attributes (see the next section in this chapter).

Alongside being able to communicate how your dissertation exemplifies specific skills or attributes, think about how to articulate how it illustrates your interest in or passion for a particular topic or issue. It is sensible to have distilled your dissertation into a clear and concise summary – this may be as a short, written document (see later in this chapter) but also as a verbal 'elevator pitch'.

DOI: 10.4324/9781003584315-15

An 'elevator pitch' is where you can summarise your idea (in this case your dissertation) in 2 or 3 minutes for a potential employer or investor. If you are asked as part of a job interview to briefly explain your dissertation, the panel are looking for your ability to condense and clearly communicate the focus, findings, and importance of your dissertation. To do this successfully, you need to think carefully about the key information you need to communicate that will allow a stranger to understand what your dissertation was about, how you did the research, what the key findings were, and why they matter. You can find useful online guidance on developing elevator pitches from many university careers centres as well as recruitment and marketing companies.

MAPPING YOUR SKILLS

Completing the skills mapping process (a process you should repeat to identify non-dissertation related examples as well) will help you prioritise how and where to use your dissertation to evidence your suitability for the advertised role. This might be in relation to being able to articulate specific technical or research skills, or for broader motivational questions. These kinds of questions may request examples of your time or project management, of overcoming adversity or unanticipated challenges, or in communicating clearly with different stakeholders.

To help answer these kinds of questions, you can think about what skills you have you developed or demonstrated during your dissertation, and then translate these into the language used by employers. Your university may have a particular framework it uses – and encourages you to use – to understand and communicate these skills: at my institution, this is the Sheffield Graduate Attributes (SGA) framework comprising 36 skills or attributes that are grouped into 12 main attribute categories:

- Academic skills
- Applying knowledge
- Research and critical thinking
- Digital capability
- Interpersonal skills
- Working with others
- Equality and inclusion

- Ethics and sustainability
- Positive well-being
- Purpose
- Personal development
- Enterprising

Taking inspiration from the relatively abstract categories within this rubric we can begin to see how you might use your dissertation to demonstrate various of these attributes. Let's start with *academic skills* – encompassing reading, note taking (and organisation), writing, data management, and data presentation. Whatever the focus of your dissertation or the methods used, you will utilise and evidence these skills: from identifying and using new sources and information, through the process of taking and organising notes from existing literature, to writing your dissertation and communicating your understanding and argument clearly and succinctly. Your dissertation then also evidences a range of *communication and interpersonal skills* – from writing, through discussions about and your ability to defend your thinking in supervision meetings, to interactions with gatekeepers, stakeholders, and participants in your research. It is likely that you have drawn on ideas from other parts of your degree programme in developing your dissertation, demonstrating you can *translate knowledge and skills* to different activities, utilising this expertise to address questions or challenges elsewhere. To do this effectively, you will also have been using *critical thinking* skills to assess the appropriateness of these ideas, and then to ask questions of existing literature and in analysing your data and findings.

Unless you are writing a purely theoretical dissertation, you will also be *handling and analysing data* – whether these data are primary or secondary, qualitative or quantitative – to inform and evidence your argument and conclusions. You will, therefore, also have had to utilise a range of *research skills* in developing and implementing a research design, identifying necessary data, data sources. methods, and analytical tools to use. It is quite likely that at some stage in the dissertation process you encountered a problem or challenge – perhaps the planned methods proved ineffective or you couldn't access a key dataset – that you had to find a solution or work-around for. Articulating how you have adapted your plans and been creative in *problem-solving* also allows you to talk about further attributes, including perseverance, resilience, creativity, and reflective personal development. You've used quantitative data in your dissertation? This demonstrates *numeracy* skills and the ability to appropriately calculate, analyse, and present numerical data.

A key challenge with these higher-level framings of skills can be the translation of these ideas into more specific and concrete examples or categories that can be mapped onto the 'essential' and 'desirable' attributes listed in job adverts. To help with this process, map specific skills and evidence of these in relation to research and technical skills, transferable skills, and personal skills/attributes. One way to do this can be to set up a spreadsheet or table (such as the example presented here as Table 15.1) that allows you to identify the skill/skill set, your experience/evidence, and your evaluation of your level of expertise in this (e.g., highly proficient/fluent,

Table 15.1 Skills Mapping and Reflection

Skill(set)	*Example*	*Self-Evaluation*
Research and technical		
Problem/question identification		
Research design		
Research methods – qualitative data collection		
Research methods – quantitative data collection		
Data analysis – qualitative		
Data analysis – quantitative		
Data presentation – writing		
Data presentation – visual		
Data presentation – oral		
Transferable skills		
Problem-solving		
Time and project management		
Critical thinking		
Global/social awareness		
Digital competencies		
Personal skills/attributes		
Resilience/determination		
Leadership		

(*Continued*)

Table 15.1 (Continued)

Skill(set)	*Example*	*Self-Evaluation*
Teamworking		
Ethics and integrity		
Adaptability		
Creativity		
Self-motivation		

proficient, basic). This final column can also help you to identify areas for further skills development – as part of reflective practice – and as a potential response to the interview question of 'Tell us about your weaknesses'.

REFLECTIVE PRACTICE AND YOUR DISSERTATION

It is likely you have been repeatedly encouraged to develop reflective practice during your degree. The previous section outlined how to start mapping your dissertation to key skills: this section provides guidance on how to narrate these connections in more detail.

This process starts with your willingness to honestly reflect on your experiences and identify what you planned to do, what you did, how it went and what you learnt from the experience. Consider how you might respond to a common interview question such as:

"Tell me about a time you demonstrated adaptability and resilience."
"Give me an example of when you have delivered a project to timely completion."
"In this role you will need to juggle multiple workstreams and deadlines. Please give me an example of when you have demonstrated these skills (of time and project management)."

How could you use your dissertation in responding effectively to these questions? You could talk in very general terms about how you had to attend lectures and submit other assignments while doing you dissertation, but this would feel very superficial. To develop your answer further here, you could use the STAR(R) approach and identify in more specific (but still succinct) terms examples from your dissertation experience.

THE STAR(R) APPROACH

The STAR(R) approach is a commonly used technique in responding to interview questions that allows you to show your skills but also your ability to (critically) reflect on your personal and professional development. The acronym stands for:

Situation
Task
Action
Result
Reflection

In practice, the STAR(R) approach should give you the scaffolding to communicate clearly and precisely how you have demonstrated a particular skill using a detailed and illustrative example. For instance, in response to an interview question on your experience of successfully managing competing demands and setting priorities the STAR(R) approach could help develop the following response:

S – Situation: provides the context of your answer (e.g., *during the third year of my degree, I had to complete an independent dissertation while completing four other taught modules and working part-time as a research assistant for a local charity*).

T – Task: explains what you were responsible for/trying to do (e.g., *for my dissertation I had to design and deliver a 10,000-word piece of research on community perceptions of the role of river bank rewilding and soft engineering as a flood defence mechanism in Huddersfield. I also had to submit eight pieces of coursework and complete a research report for the local charity on strategies to tackle anti-social behaviour to be presented to the city council*).

A – Action: what you did to successfully deliver the task(s) (e.g., *these competing demands meant I had to employ time and project management strategies, including deadline mapping and time allocation for each activity based upon interim and final deadlines, lecture times, and other commitments. For each project I identified both the priority needs in relation to deliverables and the sequencing of activities*).

R – Result: what the outcome was (e.g., *through these time management and prioritisation strategies, I was able to meet all of the deadlines in full and deliver high-quality materials – my coursework consistently secured high grades, and my report to the city council has been cited in their new 'Tacking anti-social behaviour strategy'. My dissertation has also been nominated for a Royal Geographical Society prize*).

(R) – Reflection: you don't always need to add this in, but here you reflect on how things went and whether/how you would do things differently (e.g., *While I managed my time effectively to deliver on these outcomes, I underestimated the time it took to complete the final proofreading and editing of my dissertation – and would factor in extra time for these final steps in future*).

As this notional example illustrates, your dissertation can provide ample evidence to demonstrate both specific and transferable skills and attributes to employers. You might also be able to leverage your dissertation in other ways to further support your career development and opportunities through the sharing of your findings with other audiences.

COMMUNICATING YOUR DISSERTATION TO OTHER AUDIENCES

It is useful to be able to synthesise and communicate the key messages from your dissertation to potential employers and interviewers as part of a job application and interview process. But these are not the only potential non-academic audiences for your dissertation, nor is this the only employability-related way in which you can utilise your dissertation.

The first step in this process is to think about your potential audiences and why you want to share your work with them. If your dissertation has been based on research conducted on or with particular people or organisations, you may need to share a version of your dissertation with these stakeholders. In other situations, you may have been asked by a third-party to share your findings with them. Or it may be a more speculative approach where you think your work will be of interest or relevance to an organisation –giving you a reason to contact potential future employers. Or you may be thinking about your professional profile and how you might enhance this through strategic use of online media spaces. Various options for this exist, including developing your work into a briefing note, blog post, or media release.

BRIEFING NOTE

Briefing notes set out to communicate findings and recommendations to a non-academic audience in a short and easy-to-read format. These documents need to be well designed, visually appealing, and get the main message across to the reader very quickly. These documents are typically between one and four pages in length. A briefing note will contain several key sections, whether using these specific titles or not:

- Executive summary
- Introduction/background statement
- Approach and results
- Key findings/conclusions
- (Policy) recommendations and implications

Each section needs to be clearly differentiated – often using text boxes and other visual cues. While colour and graphics can be used to great effect, they can also be distracting or even a hinderance to the reader – so think carefully and consider your intended audience. You want your briefing note to look professional and polished so avoid too many fonts or colours – and be aware of accessibility issues in terms of ensuring adequate contrast between background and font colours.

As with any piece of writing, thinking about your audience is crucial: who are they, why do they want to read your briefing note, how much do they already know/need to be told on the background subject area, how much time do they have, are your findings relevant to the audience, and what do you want them to take away from your document? Starting with these questions and a blank sheet of paper you can start to think through the three key aspects of a briefing note:

1 Audience – who are you communicating with and why?
2 Purpose – what do you want your audience to do with your findings and recommendations?
3 Message – what do you want to tell your audience? What is *the* key take-home message?

From these questions you can begin to develop and structure your briefing note, focussing on the priority message for your reader and ensuring they see this first. You can do this through placing key recommendations in a prominent position on the first page of the note, and through careful use of (sub)headings and visuals.

Briefing notes should make use of visual cues: from layout to font to (sub)headings to text hierarchy to images. Use shorter sections and informative (sub)headings, make use of clear space and text boxes to highlight sections, and don't over-complicate the presentation with multiple fonts or other unnecessary clutter. Do, though, make use of sidebars and other layout options to engage the reader, and do ensure your title (and (sub)headings) are short, catchy, and to the point. You might make use of questions to hook a reader in (Writing a good policy brief: what's the point?), use a subtitle to offer a bit more detail (Writing a good policy brief: getting your way with decision-makers), or find other strategies that work for you.

Each section of your briefing note will serve a specific purpose:

Executive summary – You should write this last but position it first in your document. This section provides a very succinct overview of the main finding or argument, the importance of the issue, and an outline of the document overall. If helpful, think of this as a super-short abstract.

Introduction/background – This provides contextual background and details the significance of the issue overall (saying why this set of findings matter), as well as an overview of the main findings.

Approach and results – This sets out how you have researched the briefing note and the key findings and analysis. You don't need to provide reams of information on methods, just a short note to indicate that the briefing note is based on a competent research design, the methods used (and how the research was done by), and evidence that it can be trusted as a reliable source. The results should then be concise summaries of findings and analysis – potentially using the occasional powerful quote or other data (in written or visual form) – moving from general to specific.

Key findings/conclusions – This section presents clear and concise conclusions and profile your key findings (a bit like in your dissertation but much more concise). These findings should be relevant to and underpin your key recommendations.

(Policy) recommendations and implications – This section sets out what you think should happen/be done, and what this would/should lead to. It is vital to ensure the recommendations proposed are relevant to the audience you are writing for.

You can find further guidance and resources on writing effective (policy) briefs by searching for the following online:

International Development Research Centre – How to write a policy brief

Policy Scotland – How to write a policy brief

University of York – How to write a policy brief

World Bank – Developing policy briefs

World Health Organisation – Policy brief template: How to write a policy brief

OTHER COMMUNICATION STRATEGIES

Depending on the topic and focus of your dissertation, there may be scope to explore other potential ways of communicating your findings and building your professional profile:

Blogs – If you already run your own blog, then consider whether you could use this to profile your dissertation findings? This will likely depend on the tone and style of your blog and whether you would want potential readers (and employers) reading your other posts. Alternatively, you may be able to write a guest piece for an external blog series.

You can find many useful online resources for writing effective blogs, including good resources from The University of Edinburgh – How to write an engaging blog,
WordPress – How to write a good blog post, and
Indeed.com – How to write a blog post

LinkedIn – If you are using LinkedIn (or similar professional social media), you might consider developing a short post that distils your dissertation's key findings and recommendations, and to highlight the skills this work exemplifies.

The Conversation and news media – Depending on the quality and focus of your work, you may want to consider pitching a very short summary of your work to an outlet such as *The Conversation* or to a local newspaper if your work would be seen as 'newsworthy'.

Good sources for tips on writing media/news releases can be sourced online from
MediaTrust,
The Guardian, and Indeed.com

CHAPTER SUMMARY

Given the amount of time and effort you will have spent on your dissertation, it is useful to understand how you can make the most of this investment. This chapter has set out some useful strategies to help you translate your dissertation into the realm of employability – including how to use the STAR(R) approach to articulate key skills and knowledge in response to interview questions. Additional options for making use of and sharing your dissertation findings with different audiences have also been outlined. While the submission of your dissertation may be the end of a key journey within your degree you can continue to utilise this work long after you have handed it in.

16

THE END OF A JOURNEY?

As you get towards the end of your dissertation journey the final push to get this work over the finishing line and submitted can feel like an uphill struggle. It might be marked by a tension between a desire just to get the document finished as quickly as possible so you never have to work on it again and a wish for more time to allow you to do more work to improve your dissertation further. Ultimately, you must hand the work in by your submission deadline – and this deadline can be both a carrot (the celebration of handing in this major piece of work) and a stick (that the work must be done by this date). While the final days or weeks of this process may feel insurmountable, it is both doable and worthwhile. Just remember, while your dissertation is important it is not your lifetime's work.

Submitting your dissertation is, though, the end of *a* journey – if not the end of *the* journey of your undergraduate degree. It is common for your dissertation to be handed in while other modules are still being taught and/or assessed. While your dissertation hand-in is likely to be a significant milestone on your degree journey it is unlikely to be the final one – and a key challenge can be recentring your energies and attention on successfully completing remaining modules and assessments.

Nonetheless, it is important to recognise the submission of your dissertation as a key milestone in your degree and a moment for celebration (however you see fit). This will have been one of – if not *the* – most challenging components of your degree. So, once you have submitted your dissertation – congratulations! You may well feel that there should be some ceremonial celebration of the moment of submission – marked by a fanfare and shower of balloons

DOI: 10.4324/9781003584315-16

and streamers. Unfortunately, the reality of the moment is more of an anticlimax – a simple clicking of the 'submit' icon on TurnItIn or similar submission platform. In most departments, the days of students descending to the office to hand-in paper copies of their work are also long gone. These did at least provide some sense of camaraderie and collective celebration (and relief).

It is important that you recognise the moment of hand-in: this is the culmination of weeks and months of work. You will have invested in this work on multiple levels and you may well feel a mixture of joy, happiness, relief, worry, and various other emotions. Allow yourself some time to process these and to mark the occasion as you see fit – in some departments, there may an organised 'hand-in party' (often by a student society) you want to attend, in others it may be informally organised among friends, or you may just want to relax (and sleep). Whatever your preference, allow yourself a moment to acknowledge success in completing this significant piece of work. If possible, have at least an evening or day away from studies to relax and rest. You need to recharge and recover – it can be a struggle to simply carry on in terms of energy and commitment and just transition to another topic/piece of work. You do, though, then need to refocus energies on final modules, lectures, assignments, and exams.

Beyond the immediate post-hand-in period, it is useful to remember that you can use the experience of doing your dissertation (and any positive feedback you get on this) in job application and interviews and to develop and advance your post-university career. Your dissertation will have helped you develop and evidence an array of skills and attributes – critical thinking, time management, project design, technical skills, communication skills, and so on. It is worth spending a bit of time soon after completing your dissertation to map these skills and experiences against commonly sought-after graduate attributes (see Chapter 15: *What next? Employability and your dissertation*). You may also find dissertation is considered for a prize – either at departmental level (if offered) or nominated for one of the many Royal Geographical Society (with Institute of British Geographers) research group dissertation prizes offered each year – if so include this in your CV as it is a marker of the quality of your work. You will have worked closely with your dissertation supervisor during this journey: they will have developed insights into your academic skills, transferable skills, and professionalism. Many students maintain some degree of contact with their supervisors for a few years post-graduation as supervisors can be very helpful in providing references for future job or further study applications.

Hopefully you will have found your dissertation journey to be a positive and interesting one, allowing you the chance to develop detailed insights and understanding of a specific issue or topic that is of significant interest to you. Whatever your experience has been, congratulations on getting to the end of it and submitting your dissertation. Hopefully this book has been of use during the writing phase of this project – and continues to be of use in translating that hard work into a resource to use as you move on from your undergraduate degree and to the next chapter of your life.

REFERENCES

Abrahamsen, R. & Williams, M.C. (2007). Securing the city: private security companies and non-state authority in global governance. *International Relations*, 21, 237–253. https://doi.org/10.1177/0047117807077006

Ah Goo, D., Hammett, D., & Hoogendoorn, G. (2025). "Woodstock was a 'lekka' place": Neighbourhood gentrification and generic-ification in a South African city. *Urban Geography*, 1–7. https://doi.org/10.1080/02723638.2025.2518264

Anderson, E. (2004). The cosmopolitan canopy. *The Annals of the American Academy of Political and Social Science*, 595(1), 14–31.

Boundary Commission for England. (2023). *2023 Review Final Recommendations Map*. Available at https://boundarycommissionforengland.independent.gov.uk/interactive-map/. Accessed 19th December 2025

Climate Data. (2025). *New Delhi Climate*. Available at https://en.climate-data.org/asia/india/delhi/new-delhi-30/. Accessed 19th December 2025

Ding, Z., Wang, J., Song, Y., Zheng, X., He, G., Chen, X., Zhang, T., Lee, W.-J., & Song, J. (2025). Tracking the carbon footprint of global generative artificial intelligence. *The Innovation*, 6(5), 100866, https://doi.org/10.1016/j.xinn.2025.100866

Dookie, D. (2024). *Essential role of climate information systems for early action interventions and resilience-focused decision-making*. ODI/RESI Working Paper. Available at https://odi.org/en/publications/role-of-climate-information-systems/

Ehrkamp, P., & Nagel, C. (2017). Policing the borders of church and societal membership: immigration and faith-based communities in the US South. *Territory, Politics, Governance*, *5*(3), 318–331. https://doi.org/10.1080/21622671.2017.1290545

Eurostat. (2025). *Land Cover Statistics*. Available at https://ec.europa.eu/eurostat/statistics-explained/index.php?title=Land_cover_statistics#Source_data_for_tables_and_graphs. Accessed 9th May 2025.

Field, R. (2016). Exploring and presenting quantitative data, in N. Clifford, M. Cope, T. Gillespie, S. French (eds) *Key Methods in Geography*, Sage, London, pp. 550–580.

Fox, A., Ziervogel, G., & Scheba, S. (2021). 'Strengthening community-based adaptation for urban transformation: managing flood risk in informal settlements in Cape Town'. *Local Environment*, 28(7): 837–851. https://doi.org/10.1080/13549839.2021.1923000

Goodfellow, T. (2022). *Politics and the urban frontier: Transformation and divergence in late urbanizing East Africa*. Oxford: Oxford University Press.

Hammett, D. (2007). *Constructing Ambiguous Identities: Negotiating Race, Respect, and Social Change in 'Coloured' Schools in Cape Town, South Africa*. PhD Thesis, University of Edinburgh, accessed at http://hdl.handle.net/1842/2084

Hammett, D. (2008). Disrespecting teacher: The decline in social standing of teachers in Cape Town, South Africa. *International Journal of Educational Development*, 28(3), 340–347. https://doi.org/10.1016/j.ijedudev.2007.05.007

Hammett, D., Hoogendoorn, G. & Masutha, M. (2025). 'Whispered in corridors': Intra-national politics and practices of knowledge production in South African Human Geography. *Transactions of the Institute of British Geographers*, 49, e12640. https://doi.org/10.1111/tran.12640

Hammett, D., Jackson, L., & Bramley, R. (2022). Beyond 'do no harm'? On the need for a dynamic approach to research ethics. *Area*, 54(4), 582–590. https://doi.org/10.1111%2Farea.12795

Hammett, D., Martin, L., & Nwankwọ, I. (2023). *Humour and Politics in Africa: Beyond Resistance*. Bristol University Press, Bristol.

Heikkilä, M. (2023). Making an image with generative AI uses as much energy as charging your phone, *MIT Technology Review*. https://www.technologyreview.com/2023/12/01/1084189/making-an-image-with-generative-ai-uses-as-much-energy-as-charging-your-phone

Horton, H. (2024). More than 6m homes at risk of flooding in England, says Environment Agency. *The Guardian*. 17th December 2024, https://www.theguardian.com/environment/2024/dec/17/more-than-6m-homes-at-risk-of-flooding-in-england-says-environment-agency. accessed on 17th December 2025

Horton, H. (2025). Dartmoor wild boar sightings prompt suspicions of guerrilla rewilding. *The Guardian*, 22nd February 2025, https://www.theguardian.com/environment/2025/feb/25/dartmoor-wild-boar-sightings-prompt-suspicions-of-guerrilla-rewilding. accessed 17th December 2025

IPCC. (2021). *IPCC Sixth Assessment Report*. Available at https://www.ipcc.ch/report/ar6/wg1/ Accessed 17th December 2025

Kraftl, P., Lynch, I., Jarman, P., Menzel, A., Walker, A., Till, R., & Hadfield-Hill, S. (2021). So you're literally taking the piss?! Critically analysing and accounting for ethics (and risk) in interdisciplinary research on children and plastics. *Children's Geographies*, 1–16. https://doi.org/10.1080/14733285.2021.1875124

Langdon, P. G., Leng, M. J., Holmes, N., & Caseldine, C. J. (2010). Lacustrine evidence of early-Holocene environmental change in northern Iceland: a multiproxy palaeoecology and stable isotope study. *The Holocene*, 20(2), 205–214. https://doi.org/10.1177/0959683609354301

Leslie, R. (2024). *Quantification of solubility trapping in natural CO_2 reservoirs using gas geochemistry and reservoir modelling*. PhD Thesis, School of Geoscience, University of Edinburgh.

Magaya, S., & Fitchett, J. M. (2022). Approaching positionality in research on indigenous knowledge systems, in Ebhuoma, E.E., & Leonard, L. (eds) *Indigenous Knowledge and Climate Governance: A Sub-Saharan African Perspective*. Cham: Springer International Publishing, pp. 81–93.

Montello, D. R., & Sutton, P. (2012). *An Introduction to Scientific Research Methods in Geography and Environmental Studies*. Sage, London.

Myers, P., & Smith, D. (2010). As football begins, West is warned: don't stereotype whole of Africa. *The Guardian*, 11 January 2010.

Nagel, C. R. (2009). Rethinking geographies of assimilation. *The Professional Geographer*, 61(3), 400–407. https://doi.org/10.1080/00330120902941753

Nicholson, D. T., Harrison, M. E., & Whalley, W. B. (2010). Assessment criteria and standards of the Geography Dissertation in the UK. *Planet*, 23(1), 18–24. https://doi.org/10.11120/plan.2010.00230018

Parsons, T., & Knight, P. (2015). *How To Do Your Dissertation in Geography and Related Disciplines*. Routledge, Abdingdon.

Pering, T. D. (2026). *Exploring Data & Statistics in the Social Sciences*. Sage, London.

Peters, K. (2017). *Your Human Geography Dissertation: Designing, Doing, Delivering*. Sage, London.

Provost, G. (2019). *100 Ways to Improve Your Writing: Proven Professional Techniques for Writing with Style and Power*. Penguin Random House, New York.

Provost, G. (2020). *Make Your Words Work: Proven Techniques for Effective Writing for Fiction and Non-fiction*. Crossroad Press, Hertford, NC.

Raento, P. (2009). Tourism, nation and the postage stamp: Examples from Finland. *Annals of Tourism Research* 36(1), 124–148. https://doi.org/10.1016/j.annals.2008.10.006

RGS-IBG. (2024). *Responsible and Ethical Fieldwork: Working in/on an Environment*. https://www.rgs.org/research/higher-education-resources/ethical-fieldwork-working-on-environment. Accessed 11th December 2024

Rieger, J., Devlieger, P., Van Assche, K., & Strickfaden, M. (2022). Doing embodied mapping/s: Becoming-with in qualitative inquiry. *International Journal of Qualitative Methods*, 21. https://doi.org/10.1177/16094069221137490

Russell Group. (2023). *Principles on the Use of Generative AI Tools in Education* https://www.russellgroup.ac.uk/policy/policy-briefings/principles-use-generative-ai-tools-education

Steinberg, P., & Peters, K. (2015). Wet ontologies, fluid spaces: Giving depth to volume through oceanic thinking. *Environment and Planning D: Society and Space*, 33(2), 247–264. https://doi.org/10.1068/d14148p

Strunk, W. (2006). *The Elements of Style*. Dover Publications, New York.

Swales, J. (1990). *Genre Analysis*. Cambridge University Press, Cambridge.

Sword, H. (2012). *Stylish Academic Writing*. Harvard University Press, Cambridge, MA.

The World Bank. (2024). *World Development Indicators*. Washington, DC: The World Bank. Available from https://datacatalog.worldbank.org/search/dataset/0037712. Accessed 17th November 2024

Truss, L. (2003). *Eats, Shoots and Leaves: The Zero-tolerance Approach to Punctuation*. Fourth Estate, London.

UN. (2025). *United Nations Development Report*. Available from https://hdr.undp.org/data-center/human-development-index#/indicies/HDI. accessed 9th May 2025.

South African Weather Service. (2025). *Historical Rain*. Available at https://www.weathersa.co.za/home/historicalrain. Accessed on 19th December 2025

Vevers, D. (2025). Illegal release of four lynx 'could be start of rogue rewilding campaign'. *The Daily Record*, 15 January 2025, https://www.dailyrecord.co.uk/news/scottish-news/illegal-release-four-lynx-could-34480385. Accessed 19th December 2025

Wallach, Y. (2011). Creating a country through currency and stamps: State symbols and nation-building in British-ruled Palestine. *Nations and Nationalism* 17(1), 129–147. https://doi.org/10.1111/j.1469-8129.2010.00470.x

Willis, J., & Todorov, A. (2006). First impressions: Making up your mind after a 100-ms exposure to a face. *Psychological Science*, 17(7), 592–598. https://doi.org/10.1111/j.1467-9280.2006.01750.x

World Bank. (2023). *Urban Development* https://www.worldbank.org/en/topic/urbandevelopment/overview. Accessed 25th May 2023

Zewe, A. (2025). Explained: Generative AI's environmental impact. *MIT News* https://news.mit.edu/2025/explained-generative-ai-environmental-impact-0117

INDEX

Pages in *italics* refer to figures and pages in **bold** refer to tables.

For Product Safety Concerns and Information please contact our EU representative GPSR@taylorandfrancis.com Taylor & Francis Verlag GmbH, Kaufingerstraße 24, 80331 München, Germany